T0315535

Practical Risk Management for EPC/Design-Build Projects

Practical Risk Management for EPC/Design-Build Projects

Manage Risks Effectively – Stop the Losses

Walter A. Salmon
FRICS MAPM MACostE

This edition first published 2020
© 2020 John Wiley & Sons Ltd

All rights reserved. No part of this publication may be reproduced, stored in a retrieval system, or transmitted, in any form or by any means, electronic, mechanical, photocopying, recording or otherwise, except as permitted by law. Advice on how to obtain permission to reuse material from this title is available at http://www.wiley.com/go/permissions.

The right of Walter A. Salmon to be identified as the author of this work has been asserted in accordance with law.

Registered Offices
John Wiley & Sons, Inc., 111 River Street, Hoboken, NJ 07030, USA
John Wiley & Sons Ltd, The Atrium, Southern Gate, Chichester, West Sussex, PO19 8SQ, UK

Editorial Office
9600 Garsington Road, Oxford, OX4 2DQ, UK

For details of our global editorial offices, customer services, and more information about Wiley products visit us at www.wiley.com.

Wiley also publishes its books in a variety of electronic formats and by print-on-demand. Some content that appears in standard print versions of this book may not be available in other formats.

Limit of Liability/Disclaimer of Warranty
While the publisher and authors have used their best efforts in preparing this work, they make no representations or warranties with respect to the accuracy or completeness of the contents of this work and specifically disclaim all warranties, including without limitation any implied warranties of merchantability or fitness for a particular purpose. No warranty may be created or extended by sales representatives, written sales materials or promotional statements for this work. The fact that an organization, website, or product is referred to in this work as a citation and/or potential source of further information does not mean that the publisher and authors endorse the information or services the organization, website, or product may provide or recommendations it may make. This work is sold with the understanding that the publisher is not engaged in rendering professional services. The advice and strategies contained herein may not be suitable for your situation. You should consult with a specialist where appropriate. Further, readers should be aware that websites listed in this work may have changed or disappeared between when this work was written and when it is read. Neither the publisher nor authors shall be liable for any loss of profit or any other commercial damages, including but not limited to special, incidental, consequential, or other damages.

Library of Congress Cataloging-in-Publication Data applied for

ISBN HB: 9781119596172

Cover Design: Wiley
Cover Image: © krisanapong detraphiphat/Getty Images

Set in 10/12pt WarnockPro by SPi Global, Chennai, India

10 9 8 7 6 5 4 3 2 1

Contents

List of Figures:

Attachments:

Appendices:

Foreword

Within any jurisdiction there are cyclical deviations in the type of and investment in major projects that are driven by investment priorities, political mandates and corporate strategies. However, at a global level there has, and will continue to be, huge investment. In 2018, the investment in major projects was around $4 trillion and, by all accounts, this investment will increase significantly year on year in an attempt to service the world's rapidly expanding population.

The procurement method on these projects varies but, in the main, the risk of both design and construction is passed to the contracting team either through a design and build, EPC or an EPC variant framework. That is all well and good, but the reality of the procurement process is that there is often a total disconnect between those that draft the contract and those that deliver the project. The contract is the single most important document on a project and in theory creates a platform of certainty, allowing those parties to it to organise and plan. However, so much of the focus in contracts is on risk allocation and risk consequence. If one asks lawyers the sorts of things that are most important on a major project, they will say things such as limitation of liability, price, termination rights and so on. However, if one asks the project team what things are most important, they may say things like reaching and maintaining a consensus around scope, having clarity as to the responsibility of the parties and good communication lines. In other words, in the eyes of the project team, what drives success in projects is about relationships and underlying governance, not management of risk consequence.

I know of a number of very good books on the subject of EPC contracts and major projects, but by and large these are written by lawyers for lawyers, and so the focus tends to be on the sorts of legal issues I identified above, sometimes with practical advice obtained vicariously. The unique offering of this book is that it purposefully focuses away from contract interpretation and legalese towards navigating the reader through a path to understanding the issues and solutions on live projects from the perspective of someone who has been there and seen it many times over. The author's experience, working for over 40 years at the centre of major projects across various sectors for some of the world's biggest companies, has been methodically and logically distilled into a treasure chest of wisdom. The book clearly sets out his understanding and interpretation of risk and putative problems. It explains in great detail all of the key roles on major projects and how those roles can and should interface successfully. Throughout the chapters, it offers practical tips and guidance for navigating a route to a successful project, or placing the contractor in the best possible position when problems arise.

The book will be of great use to contractors, project managers, employers and lawyers. It provides insight for those responsible for managing projects, those who wish to properly understand the various roles, interfaces and risks that need to be considered, and to lawyers who wish to gain a better understanding of how major projects operate and what adjustments to drafting might be taken, or live project advice might be given, to avoid or mitigate risk. If even part of the guidance given in this book is put into practice, I have little doubt that the incidence of disputes would dramatically reduce, as would the negative impact of risks on the contractor's profitability.

Partner – Jones Day   *James Pickavance*

Preface

The idea for me to write this book came about from the many questions I received from colleagues when I was assisting Project Managers in the compilation and updating of the Risks Registers for the primary risks particular to their Engineering, Procurement and Construction (EPC) Projects. At the same time, I was often involved in chairing the Risk Management Meetings on behalf of those Project Managers, and then following up the required actions with my colleagues. My participation in those tasks came about despite my principal role having been to undertake the contract administration functions for the mega EPC Projects I had been allocated to. This was simply because most other managers seemed to believe that risk management was more suited to my line of work than theirs. In the course of undertaking those tasks, some of my co-workers from whom input was required told me that the concept of risk management was a mystery to them, and it was therefore a subject that they shied away from. I had a fair degree of sympathy for them, largely because, when I first embarked on studying the subject of project risk management in the mid-1980s (as part of a project management course I was undertaking in pre-internet days), I was overwhelmed by the almost pure academic treatment of the subject (as compared with the practical application I had been looking for). Sadly, I find that this is still the case today, as my research on the internet has revealed.

Even sadder is the fact that, despite the initial great hopes for the EPC Project approach, many of the completed mega projects failed to make the profit level that the contractors had anticipated, and a great number of such projects ended in financial disaster for the contractors involved. My base discipline of quantity surveying, coupled with the contract/claims administration and risk management work I undertook, provided me with the ideal standpoint to see where and why things had gone wrong financially on the projects I had been directly associated with. I also observed the same problems occurring with monotonous regularity from one project to the next. My analysis of the countless loss-making issues I had to deal with led me to conclude that the vast majority of problems occurred for the same basic reason. Simply put, many people whose responsibility it had been for handling the work effectively were not fully aware of all the risks involved (or the fact that it was their direct responsibility to take care of such risks). Because of that, they had not taken steps to ensure that suitable mitigation plans were put in place to prevent those risks turning into ugly problems.

In principle, managing construction projects well is predominantly all about the contractor applying common sense, but it quite obviously is nowhere near as easy as that may sound. As I was in the throes of writing this book, news of the demise of Carillion plc came through (it went into compulsory liquidation on 15 January 2018). That undoubtedly sent shivers down the spines of a number of other large contractors across the globe. Almost certainly, it was the improper handling of the day-to-day major risks by the senior management team that brought Carillion down, and which was therefore a highly preventable outcome. In the construction industry, such disasters are rarely brought about by sudden unexpected events. I am sure that the Carillion management team applied common sense in their business, but I am equally sure that it could not have been applied 100% of the time, and certainly not to the most important risks.

Today, the term 'operational excellence' is very much in vogue, and most likely the Carillion management team also thought they had that subject cracked. However, I consider that, without attention being paid to excellence in risk management at all levels, meaningful operational excellence will remain a pipedream for most contractors. I therefore decided that it was perhaps time I put down my findings in a book, based on the experience I have gained over the last 20 years working on EPC Projects, about where the major risks lie for the EPC/Design-Build contractor, and how best to handle those risks. My primary objective was simply to try to help put the profit back into EPC and Design-Build Projects for the benefit of those contractors who are struggling with that issue.

This book is not aimed at Tier-1 contractors (because otherwise I would quite rightly be accused of trying to teach a grandmother to suck eggs). Rather, my target audience is mid-range construction companies and their individual managers (or aspiring managers, still learning the ropes) who would like to see a comprehensive overview of where the major risks lie in wait for them on EPC/Design-Build Projects. Many such contractors are way off being Tier-1 standard operators. However, I have met many contractors' directors who think their respective companies have made it to the big league when, in reality, their companies still have a long way to go before they could be considered to be leaders in their field. If any contractor is experiencing the situation where too many of the problems outlined in this book are regularly occurring in their organisation, then I would suggest that some urgent training/re-training is needed for their management teams.

The many different standard forms of contract used in the construction industry (as well as all the bespoke contracts I have seen) incorporate certain key words/terms as proper nouns. In an attempt to make the technical content of this book as easy as possible to read, I therefore chose to use those same words/terms in capitalised form, as well as add in a few extra words/terms of my own. Further, for the purpose of ensuring clarity, I tried not to use too many ad-hoc abbreviations in the body of the book (e.g. after the first appearance of the title 'Engineering Manager' it could thereafter have been replaced throughout with the initials 'EM'). Instead, due to the huge number of personnel and technical terms involved in the construction industry, I have chosen to write the words out in full in most cases. The primary exception is, of course, the term 'EPC'. The other exceptions occur in situations where a name/title repeats many times within the same passage/section or where the abbreviation is also very well known (such as 'HR' for Human Resources, 'HSE' for Health, Safety and Environment, and 'QA/QC' for Quality Assurance and Quality Control). The reasons for taking this approach were simple enough:

- When compiling the draft of this book, I invited comments from various close colleagues working in the construction industry. Almost every person opted to go first to the section that dealt with their specific discipline and, on a number of occasions, I was asked what certain of the abbreviations meant. That is when I realised that many readers may also do that and, if they are faced with a swathe of abbreviations they do not recognise, then their appreciation of the content could be somewhat diminished.

- I have also observed that abbreviations remain fresh in a reader's mind for only a short time, and it can become very irritating to have to flip back to be reminded what they mean, since it can often reduce the impact of what is being read. However, for the sake of good order, I have included in Appendix A (Abbreviation and Acronyms) a full list of the abbreviations and acronyms I have used. I have also provided, in Appendix B (Glossary), some extensive definitions of certain of the key technical and other terms used in this book.

If what I have written helps those responsible for running EPC Projects to manage the risks better and make more money for their efforts, then I will have achieved my principal objective. If I have also inspired readers to embrace the concept of project risk management more enthusiastically, that would be an added bonus, not just for me but for the construction industry at large.

Walter A. Salmon

Acknowledgements

The sterling efforts of my friend of 35+ years, Mr Allan Maxwell, are hereby acknowledged, whose input in the middle stages of my writing was invaluable, since it contributed substantially to the improvement of the content. Not only that, Mr Maxwell's positive comments and encouragement provided me with the impetus I needed to soldier on and complete the manuscript. A number of other colleagues have also been very helpful in providing comments and suggesting improvements over the two years plus it has taken me to complete the manuscript, not the least of whom is Mr Shahnawaz Aziz; my heartfelt thanks go to all of them, too numerous to mention.

My daughter, Anita Salmon, also earned my gratitude for her dedication and patience in carefully working her way through my initial drafts in order to clean up my grammar and correct other obvious errors. As a lay person, she also raised many valid questions that enabled me to ensure my content was a lot clearer than it otherwise would have been.

In conclusion, I would also like to thank Mr James Pickavance (former Partner at Eversheds Sutherland, UK) for having taken the time to review what I had written, for providing very useful comments/encouragement along the way, and for being kind enough to write the Foreword for this book.

Chapter 1

Introduction

1.1 The Book's Focus and Objectives

I feel obliged to make it clear, right from the outset, exactly what this book is about and, just as importantly, what it is not about. My primary focus is on how best to deal effectively with the risks to a Contractor's profitability that sit within the various departments engaged on 'EPC Projects' for major construction work (where EPC stands for 'Engineering, Procurement and Construction'). My usage of the term 'construction work' in the previous sentence is intended to cover building, civil engineering and industrial engineering work of every kind, as envisaged in the ICMSC's 'Global Consistency in Presenting Construction Costs'.[1] However, I have not attempted to distinguish in this book between different types of construction work or industries, since I consider that the same basic risks apply to all EPC Projects (and also to Design-Build Projects).

Within this book I set down where I have observed poor management or outright mismanagement occurs at the manager and department levels on lump-sum EPC and Design-Build overseas projects (i.e. construction work outside the home country of the Contractors involved). Such poor performance results in large financial losses for the Contractors and, sometimes, embarrassing consequences (such as loss of reputation). This book also deals with major risks that can be imposed on projects that originate from outside the Contractor's own Project Management Team, such as the acceptance by the Contractor of onerous contractual provisions that really should have been avoided.

As the sub-title of this book indicates, my focus is very much about how Contractors can manage risks better and thereby stop certain types of losses occurring on their EPC/Design-Build Projects. I am fully aware that the materialisation of a major risk can have other dire consequences for a Contractor that go beyond money, but most other risks can be insured against or overcome by the application of good public relations efforts. The one risk that cannot be insured against in the contracting business is going bankrupt because of financial mismanagement. Of course, insurance coverage for a company's management personnel is readily available to protect them against

1 International Construction Measurement Standards Coalition (2017). *International Construction Measurement Standards – Global Consistency in Presenting Construction Costs*, 1e. International Construction Measurement Standards Coalition, p. 7.

allegations of mismanagement (through Director and Officers Liability Insurance, for example, as explained by Construction Executive).[2] However, such insurance does not offer compensation to the Contractor for the company's losses that occur through financial mismanagement.

In the course of writing this book, I received a few negative comments from some people about the 'secondary' subject matter (managing risks effectively to stop losses). They mainly agreed with the concept of advising Contractors as to how best to deal with the risks attaching to construction projects, but thought that my focus on ensuring profitability was too commercial in today's climate. I even received one comment that 'business is about much more than just making a profit', which was said as if making losses was a virtue. However, I make no excuses whatsoever for concentrating on how Contractors should make more money and ensure their profitability by managing risks better. If anybody does not believe that making a profit is critical to businesses, they should try asking the views of the stakeholders of Carillion plc, including the tens of thousands of workers who lost their jobs when Carillion's business folded. It seems that close on 30 000 suppliers and subcontractors were still owed roughly £2 billion by Carillion at the time of its collapse.[3]

I have deliberately not attempted in this book to cover the risks inherent in 'reimbursable' construction contracts (sometimes referred to as 'cost-plus' contracts). Had I not adopted that path, I believe it would very much have confused my message as to where losses occur on lump-sum EPC/Design-Build Projects. The simple fact is that many of the risks that could lead to substantial losses on lump-sum contracts are usually opportunities to make more money under the reimbursable contract situation.

I have also not attempted to deal with too many other aspects beyond managing the major *project* risks encountered by construction companies when undertaking EPC/Design-Build Projects (i.e. I have as much as possible avoided dealing with the *corporate* risks). Most certainly therefore, this book does not attempt to provide advice as to 'How Best To Run Your Construction Company'. Instead, my primary objectives in writing this book were simply to show:

(i) what I have found to be the major loss-making risks for lump-sum EPC/Design-Build Projects;

(ii) whose responsibility I consider it is for preventing those risks from materialising into problems; and

(iii) the best mitigation methods I suggest should be adopted to prevent those risks materialising.

A large proportion of the publications I have read about how Contractors should implement sound risk management dealt predominantly with corporate risks, with the subject

---

2 Construction Executive (16 April 2014). *Understanding D&O liability insurance coverage*. www
.constructionexec.com/article/understanding-do-liability-insurance-coverage (accessed 21 March 2018).
3 Strategic Risk (15 October 2018). *Carillion collapse: the lessons learnt in supply chain risk*. Newsquest
Specialist Media Limited. www.strategic-risk-europe.com/carillion-collapse-the-lessons-learnt-in-supply-
chain-risk/1428293.article (accessed 16 October 2018).

matter treated in a very generalised fashion. Most writers had not divided the subject up into digestible chunks or split the responsibilities clearly between the different Departments accountable for the required task inputs. Nearly all such publications also tended to concentrate on the various theoretical approaches available for assessing the overall risks of taking on a new Project. Very little had been written as to what the typical risks are in respect of the specific workloads of each individual Department, or how the complement of various Managers should best work together to handle those risks during Project implementation. As a result, I considered those publications were generally not specific enough to be truly useful for a Contractor's Corporate Managers, Project Managers and Department Managers. This book was therefore written in an attempt to fill the gaps and resolve the shortcomings I had observed.

It will be seen that there is a distinct lack of academic and theoretical content in this book, which was another deliberate decision I made. This was because my intent was to present a document that contained predominantly practical content that could be applied with immediate effect in the workplace on a daily basis. It is not that I object to applying academic principles to Project Risk Management – far from it. It is just that the aim of this book is to focus on practical matters. Anybody wishing to research the various theoretical approaches that can be applied to Project Risk Management can do so by following the references I have supplied within this book (and which I believe would help them to undertake further worthwhile detailed research of their own).[4]

I do not pretend that the advice in my book is intended to be a balanced document equally reflecting the interests of both parties to an EPC or Design-Build Contract (i.e. both the Employer and the Contractor). On the contrary, I have written primarily for the benefit of Contractors. I have therefore deliberately avoided mentioning situations where the Employer could take advantage of the Contractor. Having said that, nowhere do I make any suggestions that amount to the Contractor ripping off the Employer; but putting the Contractor into a legitimately stronger position, most certainly 'yes'. The purpose of my book is simply to show where managing the risks better will prevent unnecessary losses for the Contractor. Of course, nothing prevents a member of the Employer's Team from reading this book, learning where the Contractor is most likely not to be managing risks properly, and then using that knowledge to defeat the Contractor's claim for extra time and/or money. The way for Contractors to overcome that problem is for them to sharpen up their Project Risk Management capabilities.

I included the term 'Design-Build' in the title of this book, since I consider that there are only subtle differences between how EPC Projects and Design-Build Projects are set up and run. The reality is that the risks I cover in this book can apply equally to both types of Project since, under each of the different arrangements, the responsibility for the design work (as well as the procurement and construction work typical to *all* construction Projects) falls to the Contractor. I have therefore opted in later chapters not to repeat the term 'Design-Build' unless I felt it was particularly necessary to do so.

---

4 I consider that a good starting point would be the APM (Association for Project Management) publication 'Project Risk Analysis and Management Guide' (PRAM Guide).

In an effort to deal with most of the major problems that could be encountered, I have taken into consideration the worst-case contractual scenario I can think of in respect of the risks that a Contractor could possibly face in undertaking an EPC Project. That is where the Contractor is required to submit a lump-sum bid for an overseas Project in a developing country for a specialised process plant (such as an oil refinery), in the situation where the Invitation to Bid documentation issued by the Employer comprises only:

(i) an incomplete Conceptual Design;

(ii) an outline Functional Specification that requires finalisation during the bid negotiation phase (i.e. no detailed specifications are provided);

(iii) a preliminary Plot Plan (i.e. not a fixed layout) that likewise needs to be firmed up in the bid negotiation period;

(iv) a mixed complement of loosely coordinated technical and administrative requirements that contain many references to third-party standards that sometimes conflict with each other; and

(v) a set of contractual requirements that are heavily biased in the Employer's favour.

The foregoing list of inadequate documentation is a far cry from that which the International Federation of Consulting Engineers (more commonly referred to as 'FIDIC') envisages for EPC Projects (i.e. detailed, specific Employer's Requirements are expected as a norm).[5] However, I have personally experienced working on and resolving the problems for Projects where one or more of the above inadequacies occurred (and one Project where they all occurred), although I am not at liberty to identify those Projects here. I am, however, certain that such unfairly and awkwardly constructed Projects will keep appearing, just as long as there are Contractors around who are desperate for whatever work they can get their hands on in the location they wish to build their future in.

I have had a number of people who have read some of my observations of where things went wrong on construction projects say words to the effect of 'it's just bad management', as if telling people to employ 'good management' would have miraculously cured the problems I encountered. My stance is that the many people I observed suffering the negative effects of their own poor management were usually completely unaware of what they had done wrong (or had failed to do correctly). Knowing that you have to employ good management is very different from knowing what good management is. The purpose of this book is therefore to divulge what I myself have seen go wrong, and to offer my advice as to how such situations could have been handled better. My hope is that this will lead to sound Project Risk Management being put into practice more often.

1.2 The Book's Content and Structure

I wish to stress that the core content of this book is to be found in Chapter 6 (Project Roles, Functions and Responsibilities), which sets out, over the course of 26 sections,

---

5 Fédération Internationale des Ingénieurs-Conseils (FIDIC) (2017). *Conditions of Contract for EPC Turnkey Projects*, 2e. Fédération Internationale des Ingénieurs-Conseils/International Federation of Consulting Engineers.

what I consider the Contractor's key management personnel must do (or avoid doing) in order to ensure the success of a Project. However, before launching into the detailed nitty-gritty of Project Risk Management for EPC Projects at a practical level, I feel that there are a number of important matters that need to be aired and/or straightened out first. This is largely because I have received a lot of negative comments over the years from people in the construction industry about the EPC/Design-Build concept of Project implementation. Very often, such comments came from those who had little or no experience of what is involved, so their negative comments were perhaps not surprising.

The principal negative comments against the EPC approach I received were as follows:

1. Too many Contractors working on EPC Projects have failed miserably to achieve the time and cost objectives. As McKinsey & Company found in 2017, Project durations have often been horribly extended and the final costs have quite regularly far exceeded the original budgets.[6]

   (This does not appear to be as big an issue for Design-Build Projects, perhaps because their application is often to much smaller undertakings (private residences being an example). Such Projects are also often carried out by highly specialised teams with tried-and-tested technology [for example, cold storage building contractors, swimming pool installers and prefabricated building suppliers]. In such cases, the risk sharing between the Contractor and the Employer seems to be much more evenly balanced than on EPC Projects [especially when compared to the fixed-price, major turnkey ones], as was observed by Banik and Hannan.[7])

2. The Contractor is responsible for and has complete control over at least the detailed design work, and may therefore be tempted to skimp on the quality of the finished facility wherever it seems possible to get away with doing that.

3. The Employer is not able to exercise control as to which Subcontractors or Vendors are selected for the Project. This is contrary to what is often done under the Traditional Contracting approach, where it is very common for the Employer to appoint such third parties and directly contract with them (thus giving the Employer a great deal of control over those entities).

I have to say that the foregoing concerns are not unrealistic. However, there are ways in which an Employer can exercise control over each of those issues to get what every Employer requires from a completed construction development of *any* size, namely:

(i) safe completion of the Project on time and within budget, and

(ii) a first-class facility that fully meets the functional, operational, performance, reliability, availability, maintainability, and safety requirements.

---

6 Banaszak J., Palter R. and Parsons M. (March 2017). *Stopping the insanity: Three ways to improve contractor-owner relationships on capital Projects.* www.mckinsey.com/industries/capital-Projects-and-infrastructure/our-insights (accessed January 2018).

7 Banik G.C. and Hannan F. (2008). *Specialty Contractors' Perspectives on Risk Importance and Allocation of Design-Build Contracts – Abstract* (School of Architecture, CET and Construction Southern Polytechnic State University).

To demonstrate that an EPC/Design-Build approach is every bit as worthwhile and valid from an Employer's point of view as the Traditional Contracting route, I have therefore included extensive notes about the traditional procurement approach to Project implementation, and contrasted those with both the EPC and Design-Build procurement routes. These details can be found in Chapter 2 (Construction Project Implementation Routes) and Chapter 3 (EPC Project Risk Management Overview). I have also provided a substantial amount of background detail about EPC Projects in particular: both Chapter 4 (EPC Project Pre-Implementation Problems) and Chapter 5 (Overseas EPC Project Preparatory Work) deal with different aspects of this topic.

For those who consider that they fully understand the ins and outs of the different contracting arrangements covered therein, Chapter 2 may be considered to be not so useful, and some readers may even choose to skip directly to Chapter 6. Nonetheless, I venture to suggest that Chapters 3, 4 and 5 are worth a read by anybody responsible for putting bids together, or who wishes to know where things can often go very wrong before implementation of the Project begins to get underway.

My approach in Chapter 6 has been to pinpoint, from my standpoint, the appropriate Manager to whom the responsibility for identifying and managing each specific risk should rightly fall, in line with the thinking displayed by the Association for Project Management (APM).[8] I have therefore written this book as if there is only one way to organise the corporate structure and allocate Project responsibilities within a construction company. No doubt that will make my work subject to a certain amount of criticism, notwithstanding that I took that approach in full recognition of the fact that there are many different schools of thought as to how construction companies and Projects should be organised. My thinking was that, no matter how a company organises its Corporate Management Team or deals with the management of its Projects, it is not possible to remove either the corporate risks or the project risks; they will always remain, and the only things that will change from one company to another are:

   (i) upon whom the responsibility for managing each risk falls,

   (ii) how the importance of each risk is perceived, and

   (iii) how each risk is to be handled.

I also considered that, if there is no clear organisational framework or management structure established, then the principles of Project Risk Management I mention could be deemed to apply to different roles/functions within the organisation. However, I intended to pin down with a greater degree of certainty what the risks are and which person is properly responsible for managing each risk (in my opinion, of course). Thereafter, having seen how I have applied the risk management principles to the preferred structure I have chosen to follow, others with differing views can then adapt those principles to accord with their own preferences for how their companies and Projects should be organised. In order to make it clear as to what management structure I am working to, Appendix C shows the 'EPC Project Management Team Organisation Structure' I

---

8 Association for Project Management. *APM PRAM Guide*. Association for Project Management, p. 98: 'Risks are more likely to be acted on if responsibility is allocated to individuals'.

have employed and Appendix D shows the 'EPC Project Departmental Organisation Structure' I have adopted.

To round out and complete the subject of Project Risk Management, I have also included three further Chapters, namely: Chapter 7 (Reducing Joint-Venture/Consortium Risks), Chapter 8 (Claims Management Risks and Problems) and Chapter 9 (Identifying Hazards and Managing the Risks).

I had intended that construction industry students/newcomers should also be able to find the content of this book useful. In view of that, I am compelled to add the caution that the management structure I have adopted herein (for both the EPC Project team and the Contractor's company) must *not* be considered to be written in stone. It simply provided a means for me to allocate roles and functions/responsibilities in a structured manner, in an effort to ensure that I did not miss out discussing any key area of management. It therefore needs to be recognised from the outset that, within different companies, some of the functions/responsibilities I have allocated to a particular Manager may well be allocated to another Manager. Nonetheless, my hope is that my approach will enable those who have not had experience occupying or dealing with certain of the various management positions to be able to appreciate more fully what it takes to run an EPC Project efficiently and effectively. Ultimately, if all this book achieves is to encourage more discussion about the practical application of Project Risk Management in the construction industry (and for EPC and Design-Build Projects in particular) then I will consider that I have done some good, regardless of any flak that may come my way.

1.3 Generality of Contractual Advice Given

There are a number of standard forms of contract that deal specifically with lump-sum EPC/Design-Build Projects. For example:

- the FIDIC Silver Book ('Conditions of Contract for EPC/Turnkey Projects'),[9]
- the FIDIC Yellow Book ('Conditions of Contract for Plant and Design-Build'),[10]
- the 'ICE Conditions of Contract Design and Construct',[11] and
- the IChemE 'Lump Sum Contract – The Red Book'.[12]

However, International Oil and Gas Companies (such as BP, ExxonMobil, Gazprom, Lukoil, Petronas, Royal Dutch Shell and the like) commonly use (and will no doubt continue to use) bespoke contracts that have been tailored to the individual needs of each respective company. The use of such custom-compiled contracts is primarily because the standard forms of contract would need to be substantially modified to incorporate all the special requirements particular to each of those organisations.

In view of the above, the advice I give in this book has not been developed with reference to any specific standard forms of contract. In turn, that means that I have not attempted

---

9 FIDIC – Fédération Internationale des Ingénieurs-Conseils.
10 Ibid.
11 Published by ICE Publishing (part of the Institution of Civil Engineers).
12 Published by the Institution of Chemical Engineers.

to deal with any of the particular terms and conditions that are found in the standard forms. I prefer to leave that to lawyers and, in any case, those topics are covered more than admirably in a number of well-known books.[13] However, I have tried to identify as many as possible of the uglier (unfair/one-sided) clauses that work their way into many contracts, whether the base document is one of the standard forms of contract or a bespoke contract.

1.4 Common Elements for Construction Projects

Construction work is ever present amongst us, whichever direction you care to look, taking on many different forms. Construction activities range from major city redevelopments to petrochemical plants, and with so many things in between; the list of possible construction work seems endless. And the Construction Industry is simply huge. For infrastructure work alone, according to McKinsey & Company the value anticipated to be undertaken worldwide over the period 2018–2023 is in the order of USD 77 trillion.[14]

There are many steps that need to be taken first before even one spade can strike the ground in earnest to commence the physical on-site work for a major Project. For most Projects, a feasibility study will need to be conducted before the Employer will be prepared to commit the necessary resources to proceeding with the implementation activities. A typical feasibility study will address the questions of the legality of the proposed Project and whether its construction would be technically feasible, as well as being fully justified on economic grounds. It is not usual for a Contractor to be involved in ascertaining the viability and feasibility of a Project unless the Contractor is required to be involved in organising and/or providing the finances. This requirement can sometimes be necessary under an EPC Contract if payment to the Contractor is not due to be made until a number of years after the facility has been operating. It is most certainly required in Projects where the Build-Operate-Transfer (BOT) concept is employed.[15] BOT Projects may also be implemented under an EPC arrangement. The problems for some Projects can start with an inadequate feasibility study, which can then lead to severe cash flow problems for the Employer (and, ultimately, the Contractor). To avoid the possibly disastrous problems that an inadequate feasibility study can cause, it has been proposed by some (see Hyari and Kandil) that a series of peer reviews should be conducted of all feasibility study material.[16]

There are many different ways for an Employer who wants to have construction work carried out to arrange for the appointment of a Contractor to do the work.

---

13 Such as, but by no means limited to, *Hudson's Building and Engineering Contracts* (by Hudson A.A. and Wallace I.N.D.) and also *Construction Contracts - Law and Management* (by Hughes W., Champion R. and Murdoch J.).

14 Billows J., Kroll K., Pikul P. et al. (August 2018). *Capital Project value improvement in the 21st century: Trillions of dollars in the offing.* www.mckinsey.com/industries/capital-Projects-and-infrastructure/our-insights/capital-Project-value-improvement-in-the-21st-century-trillions-of-dollars-in-the-offing (accessed 05 September 2018).

15 MBA Knowledge Base (5 March 2012). *Build Operate Transfer (BOT) Model.* www.mbaknol.com/Project-management/build-operate-transfer-bot-model (accessed 31 July 2018).

16 Hyari K. and Kandil A (2009). *Validity of Feasibility Studies for Infrastructure Construction Projects.* *Jordan Journal of Civil Engineering*, 3(1).

However, whether the work is let under a Traditional Contracting arrangement or an EPC/Design-Build arrangement, the Contractor's work will almost certainly involve both procurement and installation/construction work. It will therefore be of no surprise that a great deal of what I have written will quite obviously also apply equally well to non-EPC/Design-Build Contractors, despite the intended focus of this book being on EPC/Design-Build Projects.

Whether or not a Project is conducted under an EPC or Design-Build arrangement, failure to manage the implementation risks properly for a major Project can lead not only to huge financial losses but also to bankruptcy for the Contractor's entire business. Mismanaging a major Project's risk portfolio should therefore be viewed as a gamble too far for most Contractors, and advance planning is therefore vital. Consequently, no matter what type of contractual arrangement the successful bidder will eventually be working under (whether the Traditional Contracting approach or an EPC/Design-Build route), there are certain basic preparatory bidding steps that need to be taken. The following sets out what those steps are, and which are essential if the bid pricing is to stand a good chance of adequately covering all the costs involved for undertaking all the work necessary to complete the Project successfully:

1. Planning how the work will be done through the development of a comprehensive Work Breakdown Structure (WBS) for the Project. Sometimes the starting point might be an outline WBS required/prepared by the Employer's Team and issued with the Invitation to Bid documentation. No matter, the WBS needs to be worked up into a truly meaningful list of all the work activities/elements involved and, under an EPC or Design-Build Project, the Engineering, Procurement and Construction components should each have their own list of work activities/elements clearly identified under those specific headings.

2. Using the completed WBS, the Contractor must then establish the sequence of undertaking the work activities/elements involved, and also determine the length of time needed to deliver the completed work for the entire Project (the Project Schedule). If it is also possible to establish the labour and construction equipment resources required reasonably accurately, then that would be of great benefit, since it would help to give more confidence in both the anticipated Project completion time and the Contractor's bid pricing. However, that is more often than not very difficult to achieve under an EPC/Design-Build Project, due primarily to the lack of detailed design information available from which to measure the physical work quantities that will be required for the completed Project.

3. Having prepared the WBS and established what the Project Schedule looks like, the Contractor must then conduct a preliminary risk analysis. The aim should be to establish what the major risks are that, if not controlled adequately, would have the potential to cause major problems and thereby stop the Project from being as successful as it could be. Those risks should then be set down in a preliminary Project Main Risks Register, alongside which suitable risk mitigation measures should be included (wherever it is considered possible/feasible to achieve that), aimed at preventing those risks from materialising.

4. The final essential ingredient in submitting a worthwhile, comprehensive bid is for the Contractor to prepare an outline Project Execution Plan (PEP) that incorporates all the findings from the WBS, the Project Schedule and the Project Main Risks Register. A properly prepared and well thought out PEP is, in essence, the storyline for how the Project will be undertaken. If written competently, the PEP would allow the Contractor's Project Implementation Team to form a very clear picture as to what the most effective management set-up ought to be. That too would give the Contractor added confidence about the adequacy of the bid pricing.

All of the above topics, common to all construction Projects (EPC/Design-Build or otherwise), are dealt with in far more detail in the following chapters of this book, along with advice as to (i) what to look out for when compiling the necessary information and documentation, (ii) what things can go wrong, and (iii) how to avoid such problems occurring.

Chapter 2

Construction Project Implementation Routes

2.1 Different Approaches

I freely acknowledge that there are many different ways to approach the implementation of construction Projects that will result in getting a proposed facility constructed (successfully or otherwise). However, since the focus of this book is on EPC and Design-Build Projects, I only intend to contrast those two procurement routes with what is known as the Traditional Contracting route. Also, in order to clear up any confusion that may exist in the reader's mind about the relationship between EPC Projects and the construction procurement route known as EPCM (Engineering, Procurement and Construction Management), within Section 2.5 2.5 (EPCM Approach) I also briefly touch on that particular implementation method.

2.2 Traditional Contracting Approach

2.2.1 Design Team's Appointment and Role

Having established that a Project is feasible in all respects (see Section 1.4 – Common Elements for Construction Projects), the traditional way to implement a construction Project is for the Employer to appoint a Design Team. Usually that will be either a lead architectural entity or, if more appropriate for the type of Project involved (such as, for example, a new viaduct, a sewage treatment plant or a processing facility), an engineering design entity. The role of that Design Team would be to handle the following primary tasks.

1. Undertaking the design work for the entire facility (except, perhaps, for the final working/shop drawings that need to be prepared by specialists for bespoke manufacturing or complicated fabrication and installation work). This will generally also include taking the responsibility for supervising any separate specialist engineering work (such as that done by a geotechnical survey team and structural, mechanical and electrical engineers, etc.).

2. Advising on and preparing the specifications to determine the precise quality requirements for the finished facility.

3. Advising on the most appropriate bidding strategy for selection of the Contractor and any specialist Subcontractors.

4. Working with and monitoring the work of the team preparing the bidding documents for the various work packages, including the quantity surveying team preparing the bills of quantities (if applicable). *[Note: EPC Projects will not generally have bills of quantities, due to there being no detailed design available.]*

5. Organising the entire bidding process for the construction work (including the separate bid packages for any specialist Subcontractors).

6. Advising on the bids received for the construction work and specialist subcontracted work.

7. Arranging for signing of the construction contract and subcontracted work packages by the Employer.

8. Administering the entire construction process, including:

   (i) supervising the construction work right through to completion of the commissioning activities and handover of the completed facility from the Contractor to the Employer;

   (ii) monitoring the Contractor's and Subcontractors' work quality and progress;

   (iii) issuing instructions in respect of any changes needed; and

   (iv) closing out the construction contract and the subcontracts (including advising, in conjunction with the cost engineering team and/or quantity surveying team, on the settlement of all commercial matters and any claims with the Contractor and Subcontractors).

2.2.2 Employer's Participation

Beyond appointing the Design Team, under the Traditional Contracting approach the Employer's participation usually also encompasses the following.

1. Approving the issuance of bidding documents, in which the Employer either takes on 100% of certain risks or shares those risks with the Contractor.

2. Allocating risk sharing – an example of where the Employer would usually accept the entirety of a risk is in respect of the subsoil proving to be of poor quality, leading to higher costs for the foundations. An example of a risk typically shared would be in regard to the occurrence of exceptionally bad weather, where the Contractor would be entitled to extra time for completing the work but for which no monetary compensation would be payable to the Contractor. However, the Contractor would be relieved of the responsibility for paying Liquidated Damages for delayed completion due to this occurrence.

3. Arranging separately for certain key materials, goods and/or equipment to be supplied and/or installed by entities other than the Contractor.

4. Reviewing commercial bids received from construction companies for building what the Design Team's drawings and specifications show, where the Contractor's price is often based on the bills of quantities that are provided by the Employer's Team.

5. Awarding the Project implementation work to the bidder considered to be offering the best deal to the Employer (usually the lowest price, provided that the time-frame for completion is acceptable to the Employer).

2.2.3 Contractor's Role and Responsibilities

The Contractor's role and responsibilities under the Traditional Contracting approach are as follows.

1. Preparing and submitting a commercial bid for the Project, based on the drawings and specifications prepared by the Design Team. That will most likely be against bills of quantities provided by the Employer's Team, which the Contractor can assume accurately reflect what is needed to be provided in respect of the physical work for the completed facility.

2. After award of the construction contract for the Project:

    (i) ordering the required materials, goods, and equipment in line with the specification requirements,

    (ii) providing and supervising all labour and construction equipment necessary for constructing, testing, and commissioning the completed facility, and

    (iii) handing the completed facility over to the Employer for occupation/use.

3. Rectifying, at no cost to the Employer, any defects found in the completed facility during the warranty period following handover (often referred to as the 'defects liability period').

2.2.4 Traditional Approach Advantages

It should be noted that the prime advantage of employing the Traditional Contracting approach is that it provides much more confidence about the Project being handed over by the designated completion date than with almost any other approach to construction work, provided always that a competent Contractor has been selected. This is because there is far less likely to be uncertainty about the work content, since the Detailed Design drawings will have been finalised, thereby allowing fully detailed bills of quantities to be produced. That combined level of information enables the Contractor to determine the labour and construction equipment resources very accurately, right from the time of preparing the commercial bid. The foregoing reference to the completion date is in regard only to the construction activities. It is not to say that the overall time-frame for Project completion will be quicker than with the EPC route. This is because, under the Traditional route, the Detailed Design work has to be completed before the Contractor can commence the construction activities. That is not the same case with the EPC approach.

Having a Detailed Design completed ahead of the commercial bidding phase should also ensure that delays and additional costs due to unexpected design changes in the implementation phase are very much reduced. This may well be the reason why the Traditional Contracting approach to implementing construction Projects today still has greater prevalence than any other construction Project implementation method. This can be seen by researching on the Internet to see how various governments around the world handle the implementation of construction Projects, typical of which is the South Australian government's preferred contracting strategy.[1]

2.2.5 Traditional Approach Disadvantages

Under the Traditional Contracting approach, the Employer often chooses to get involved in the direct appointment of companies to undertake key elements of the Project, such as, but certainly not limited to, curtain walling, elevators, and specialist installations (e.g. air conditioning). This means that there can be many different companies involved directly with the Employer, each requiring a separate set of contract documents to be signed with the Employer. This situation can lead to difficulties for Employers when it comes to allocating responsibility and liability to the right party for anything that goes wrong in the implementation stage or, subsequently, with the completed facility. Added to this, the Employer usually shoulders the greatest financial burden for Project delays if one of the contracting parties delays any of the other contracting parties.

Another area where the Employer is more vulnerable under the Traditional Contracting Approach is with regard to Variations, since the Contractor will be looking for reimbursement for all the extra costs involved. These extras often arise because the Contractor is very often appointed before the Design Team has fully completed the Detailed Design work; the extras creep in as the shortcomings in the design work are uncovered. By contrast, under the EPC and Design-Build approaches, once the Front-End Engineering Design has been fixed, the Contractor will thereafter be deemed to have allowed in its bid pricing for everything necessary to complete the Project. Any changes brought about due to the development of the Detailed Design by the Contractor will all be at the Contractor's expense and not be reimbursed by the Employer.

As mentioned in Section 2.2.4, one vital point about the overall relative time-frame taken to complete a Project under the Traditional Contracting arrangement needs to be remembered. This is that the time taken to compete the design work generally has to be added to the construction time to obtain the overall time-frame for handover of the Project from its conception. By contrast, under an EPC/Design-Build arrangement, a large portion of the design work can usually be undertaken while the procurement and early construction activities are ongoing. In theory at least, this means that, under both the EPC approach and the Design-Build approach, a saving can be achieved in the overall time taken to get to the point of handing over the completed Project to the Employer compared with adopting the Traditional Contracting approach.

2.3 Design-Build Approach

For those construction Projects where architectural merit may not be so important, an alternative method of implementation to following the Traditional Contracting route is the Design-Build approach. Under this approach, a Contractor is selected to both design

---

1 Government of South Australia (2005). *Construction Procurement Policy Project Implementation Process.* ISBN 0-9775044-0-7. www.dpti.sa.gov.au/__data/assets/pdf_file 0020/51653/pip.pdf (accessed 28 November 2017).

and build the facility (including selecting and appointing all the Subcontractors). This is usually conducted under the watchful eye of an Employer's representative (who may or may not be an Architect or Design Engineer). Whether or not the Contractor is given a Conceptual Design to follow will depend to a great extent on the type of facility involved, and whether or not the façade needs to be given special attention. For example, if the Employer requires a standard cold storage building to be erected quickly, then it would generally be adequate for a specialist Contractor to propose the layout and elevational details. On the other hand, if an upmarket apartment block is required, the Employer will inevitably be much more fussy about both the internal and external layout of the property, as well as its external looks, especially if the apartments are required to be sold as quickly as possible after their completion. If the latter situation is the case, then the Employer may well appoint an Architect to take care of such sensitive details, with the Detailed Design work being left to the Contractor to do.

The Design-Build approach has the benefit for the Employer that the responsibility for the functionality of the completed facility is not generally split between a Design Team and the Contractor, nor is the Employer responsible for the appointment and performance of Subcontractors. This 'one-stop shop' approach would not be achievable if the Detailed Design work were to be given to a third party to undertake (such as to a specialist designer not linked to the Contractor). The same would be true if the construction work were to be broken up between different Contractors (such as for the civil work, the buildings work or the mechanical and electrical work elements). It would also not be achievable if the procurement work were to be undertaken by the Employer (or by a third party on behalf of the Employer), since the responsibility for any materials, goods, and equipment that arrived late or failed to perform as required would not be the Contractor's responsibility. However, similar to the Traditional Contracting approach, under the Design-Build approach the Employer will usually shoulder the financial risks of such things as unexpected poor soil quality, bad weather delaying the Contractor, etc. Such risk adoption by the Employer helps to keep the bid prices down, since the Employer only has to pay extra to the Contractor if the risks actually materialise and become problems.

Since, generally speaking, the design work is not complicated for Design-Build developments, there is usually a much shortened design period compared with the Traditional Contracting route, especially if the facility is of the specialist type and the Contractor itself is the design specialist. For this reason, there is also usually no need for third-party design consultants to approve the design before the Contractor orders materials, goods and equipment, or starts the foundations work. Added to this, as already mentioned above, completion of the design work can run in parallel with the procurement and early construction activities. As the Construction Industry Institute's recent research has validated,[2] this therefore means that the time-frame for completing a facility is generally far quicker (and often cheaper) under the Design-Build arrangement than it is for the Traditional Contracting approach (where Architects and Engineers are required to complete all the design work before the Contractor selection process can begin).

---

2 Bobbitt Design Build Inc. (15 April 2019). *New study: design-build can cut project delivery time in half*. https://www.bobbitt.com/article/new-study-design-build-can-cut-project-delivery-time-in-half (accessed 6 May 2019).

The above is simply an overview of what the Design-Build approach comprises, whereas it is in fact a complex topic with a great variety of options as to who the contracting parties may be, where Architects and Engineers may play a bigger role, the different ways bidding can be conducted etc., none of which is the purpose of this book. A good starting point for those wishing to explore the concept of the Design-Build approach further is the appropriate Wikipedia article, since it contains a large number of very useful references for further reading.[3]

2.4 EPC Approach

2.4.1 EPC Project Suitability

Although the Design-Build approach can work well with simple Projects, it may not be an entirely suitable procurement route if the proposed facility is complex (with, say, a great deal of emphasis on specialised industrial engineering inputs and proprietary equipment). This is even more true if the final product must also comply with a vast amount of stringent safety regulations and environmental obligations, as well as operational and performance requirements, before being put into use. Where the Employer requires the Contractor to guarantee the quantity and quality of outputs from the completed facility in the operational phase (usually because the Contractor is required to be the entity responsible for selecting, procuring and installing the process equipment), the contractual situation becomes even more complex. Examples of such complex construction work are oil and gas pipelines, oil refineries, petrochemical plants, etc., many of which fall into the mega Project bracket (i.e. over US$1 billion).

For dealing more effectively with such complex Projects, many Employers nowadays consider it best to put as much of the risks of Project success as possible squarely on the shoulders of the Contractor, well beyond the level of risks that would usually be acceptable to a Design-Build Contractor. This is where the Engineering, Procurement, and Construction (EPC) approach comes into its own. It is far better suited for this more complex type of technical/contractual scenario from the Employer's standpoint, especially where most of the risks are to be transferred to the Contractor. The EPC arrangement also usually avoids the need for the Employer to get involved in all the many complicated interfaces and associated coordination activities that occur between the design phase, the procurement phase and the construction phase. Those interfaces can often become an unwieldy task and an awful burden (especially where utility service providers are involved). It should be noted that use of the EPC approach is not limited to complex engineering work or facilities where proprietary technology is involved. In fact, the EPC approach has also been successfully applied to many different types of construction work, including large-scale developments such as upmarket holiday resorts (often including the full fitting-out requirements, usually under what is known as a 'turnkey' arrangement).

Under the EPC approach, the Contractor is, just as under the Design-Build arrangement, still the one-stop shop for all implementation activities beyond the basic design work (including the Detailed Design work and, sometimes, even for undertaking the

---

3 Wikipedia. *Design-build*. https://en.wikipedia.org/wiki/Design-build (accessed 23 August 2018).

preceding Front-End Engineering Design [FEED] work). However, the EPC Contractor's responsibility will generally also extend to the suitability (fitness for purpose), functionality, quality, and performance of the finished facility, including the quantity and quality of its outputs (which Ron Douglas has elaborated in a whitepaper).[4] This is despite the fact that the Employer's Team will almost inevitably be far more deeply involved in approving the Contractor's work during the implementation process than under a Design-Build approach. Such deeper involvement is primarily needed to ensure that suitable 'checks and balances' are employed to confirm, as far as possible, that the completed facility can be operated safely.

Of course, the transfer of such major risks and responsibilities from the Employer to the Contractor inevitably results in higher bid prices for EPC Projects compared with using both the Design-Build and Traditional Contracting approaches. However, many Employers seem happy to settle for that trade-off, in order to gain greater certainty of the overall price to be paid (at the end of the day) and the guaranteed quality of, and outputs from, the completed facility.

2.4.2 Contractor's Obligations

Following from the above, a typical EPC Project is therefore one where the Employer appoints a Contractor to:

(i) assume responsibility for and undertake the Detailed Design work for a major facility, based on the Employer's detailed requirements (often referred to as the Basic Engineering Design or, in the Oil and Gas Industry, the Front-End Engineering Design);

(ii) procure and deliver all the necessary materials to the Project's Site location;

(iii) construct the facility in its entirety;

(iv) fully commission the facility and prepare all the operational manuals, ready for handing over to the Employer to start occupying and/or operating it immediately;

(v) guarantee the quality and quantity of the outputs (and regulated emissions) from the completed facility; and

(vi) rectify all problems found in the defects liability period, entirely at the Contractor's cost.

2.4.3 Employer's Participation

Under the EPC approach, the Employer usually does something along the following lines (although there are many different options available):

(i) appoints a Design Team (usually a highly specialised Engineering Team) to produce the preliminary design information, which could be either only the Conceptual Design and a Performance Specification or a partial or fully completed FEED, along with a Functional Specification (including full performance output requirements);

---

4 Douglas R. (2016). *EPC or EPCM contracts, which one can drive stronger outcomes for project owners?* http://www.ausenco.com/en/epc-epcm-whitepaper (accessed 11 November 2017).

(ii) issues bidding documents in which the Employer places a great deal or even all of the risks squarely on the shoulders of the Contractor;

(iii) receives technical proposals from Contractors for (a) developing the preliminary design information into the Detailed Design information, (b) undertaking the required procurement activities, and (c) constructing and commissioning the completed facility (or at least assisting the Employer's Team with the commissioning work);

(iv) assesses the technical proposals from the Contractors with a view to establishing which proposals are technically acceptable in all respects;

(v) receives commercial bids from the Contractors who had submitted acceptable technical proposals;

(vi) awards the Project implementation work to the bidder considered to be offering the best deal to the Employer (which usually means the bid with the lowest bid price); and

(vii) appoints either construction phase administrators (commonly referred to as a Project Management Consultant) to supervise the Contractor's work (and advise on such matters as variations and extensions of the time for completion of the construction work) or, alternatively, a 'representative' to monitor what the Contractor is doing (and who will not usually get too involved with the Contractor's activities, but who will act as the eyes and ears of the Employer and also as the Employer's spokesperson).

2.4.4   Standard EPC Contracts Available

Recognising the vast differences that can occur in respect of an Employer's appetite for accepting risk on large-scale design-and-install/build type Projects, the International Federation of Consulting Engineers (FIDIC) has produced two quite different standard forms of contract for Employers to choose from:

- The Yellow Book – 'Conditions of Contract for Plant & Design-Build' (2nd Edition, 2017), and

- The Silver Book – 'Conditions of Contract for EPC/Turnkey Projects' (2nd Edition, 2017).[5]

In only one of the above two FIDIC publications is the term 'EPC' applied (in the Silver Book). However, the FIDIC Yellow Book (2017) is also clearly a contract suitable for an EPC Project, since the requirement therein is also for the Contractor to complete the Engineering work as well as undertake all the Procurement and Construction activities, as is shown by its subtitle ('For Electrical & Mechanical Plant, And For Building And Engineering Works, Designed By The Contractor'). Although there is a large degree of similarity between the two sets of FIDIC design-and-install/build type contractual arrangements, one of the principal differences is that the Yellow Book shares the responsibilities and liabilities inherent in the major implementation risks more evenly

---

5 Both books published by FIDIC (also known as 'The International Federation of Consulting Engineers', FIDIC being the acronym of the French version of its name).

between the two parties (and also the consequent apportionment of cost sharing for dealing with those risks). By contrast, the Silver Book requires the Contractor to bear nearly all the risks entirely at its own expense, regardless of what those risks are or how onerous they are.

Another interesting thing to note is that the participants acting for the Employer are quite different between the FIDIC Yellow and Silver Books. In the Yellow Book it mentions the Engineer, while in the Silver Book there is only an Employer's representative (whose role is markedly dissimilar from the Engineer mentioned in the Yellow Book). The difference between the Yellow and Silver Books is primarily a result of the 'turnkey' arrangement catered for under the Silver Book, where the Contractor is expected to operate far more independently of the Employer and accept a far greater risk burden than is envisaged under the Yellow Book. Under the 'turnkey' arrangement, FIDIC expects the Employer to do nothing much more than (i) not hinder the Contractor in its implementation efforts, (ii) pay the Contractor in accordance with the contractual provisions, and (iii) wait to be handed the keys of the fully completed facility. Further, FIDIC appears to use the term 'turnkey' in the Silver Book to refer to the situation where the Employer passes over virtually every conceivable risk to the Contractor in anticipation of not paying extra for any eventualities or extra work subsequently found to be required. That arrangement is significantly different to the situation where risks are spread more evenly between the Employer and the Contractor (on the basis, say, of looking to see which party is better suited to averting/handling any given risk item), as they are under the FIDIC Yellow Book.

Where the EPC Contractor is required to take full responsibility for handing over the facility in the required time-frame based on the premise that no changes to the Employer's requirements will be evidenced, the contract is sometimes referred to as being Lump-Sum Turnkey in nature ('LSTK' being the abbreviation/initialism commonly used for that term). Under an LSTK Contract, the Employer will generally expect no extra costs to be payable to the Contractor for completing the work specified in the contract. It should be noted that all LSTK Projects are EPC Projects, but not all EPC Projects are LSTK contracts.

The choice by the Employer as to which of the above FIDIC-orientated EPC/Design-Build contractual arrangements should be adopted will depend to a great extent on the following factors:

(i) what type of organisation the Employer has and, therefore, on just how much (or how little) technical competence sits within the Employer's organisation,

(ii) how much control the Employer wishes to exercise over the implementation process, and

(iii) the price the Employer is willing to pay to pass as much as possible of the implementation risks over to the Contractor.

Some Employers prefer to adopt the equivalent of the FIDIC Yellow Book approach for EPC Projects (where an Engineer is appointed separately to represent the Employer). The Yellow Book approach is very often selected where the Employer has a competent in-house workforce to handle the supervision of the implementation activities, especially the on-site work. The alternative approach is to follow more closely the FIDIC Silver Book provisions. This is usually done where the Employer does not have adequate resources to supervise the whole of the implementation work. No separate Engineer is appointed under this alternative approach but, instead, only a representative of the Employer, whose role will be to interface with the EPC contractor (although very often subject to strict limitations of authority). Having said that, the Employer may nonetheless appoint separate companies to monitor either the engineering activities, the procurement activities, the construction activities and/or the commissioning work activities, or any combination of such monitoring activities; there are many possibilities/options available.

2.4.5 General Notes of Interest

Not all EPC Projects will require the Contractor to be responsible for commissioning the facility, and the 'turnkey' tag cannot therefore be applied if that situation applies to the Project. This will often occur where the completed facility requires hydrocarbons or special processing materials to be used that have the potential to cause harm if not handled correctly (and where specialists will often be engaged by the Employer to take responsibility for the commissioning work). This has therefore caused the term 'EPCC' to arise, where the final 'C' refers to 'Commissioning', making the whole read 'Engineering, Procurement, Construction and Commissioning'. However, this is an unnecessary appendage, since the specific requirements for each EPC Project are unique, and the scope of work definitions written into the contractual documentation will usually define the actual work content for the Contractor. In particular, the FIDIC contracts (the Silver and Yellow Books) do not make any references therein to Procurement (nor Engineering, nor Construction), thus any specific requirements for those activities must be separately specified elsewhere in the documents that form the basis of the contractual agreement between the parties. Consequently, if using either of those FIDIC publications, it is therefore sufficient to describe the contract as being for an 'EPC' Project, even where the Contractor is not required to undertake the commissioning work. The same reasoning applies to the term 'EPCI', where the 'I' refers to 'Installation' (thus 'Engineering, Procurement, Construction and Installation'); again, this distinction is unnecessary for the reasons given for not using the term 'EPCC'. Despite that, I am sure that such variations in the abbreviations used will continue to be employed widely.

In the past, many EPC Contracts were of the reimbursable type, where the Contractor was recompensed in full for all manpower and construction equipment resources used, as well as for all purchases made for the Project. However, those halcyon days for Contractors have generally long-since gone at the time of writing this book and, sadly for Contractors, may not return for many years to come. Nowadays, the tendency is for Employers to insist upon agreeing a lump-sum price with the EPC Contractor for the whole of the Project's implementation. That is an arrangement that almost always benefits the Employer much more than it does the Contractor. This book therefore focuses solely on Projects undertaken on the basis of a lump-sum payment arrangement. In addition, as mentioned in Section 1.1, this book assumes that the EPC Project is to be built in an overseas developing country, since that type of Project is far more difficult to manage than one planned to be built in the Contractor's own backyard.

Much as for the Design-Build approach, the EPC approach has many variants and is a deep subject once you decide to get into exploring its possibilities. Unfortunately however, as late as when I was completing the writing of this book (early 2019), I have to say I still found Wikipedia did not provide suitable reference material about EPC Projects. So much so, that I will not put any direct references to the Wikipedia article here, since I do not wish to waste anybody's time or, worse, for anybody to get confused. Having said that, if a Google search is conducted for the phrase 'EPC Projects', you will see that in the order of 18 000 000 references can be found by the search engine. Having myself opened and read in excess of what I thought were the best 200 or so links displayed by Google, I found that there is a lot of inadequate material on the Internet in respect of EPC Projects. Where I found good material, I have made appropriate reference to it in this book.

2.5 EPCM Approach

The 'EPC' abbreviation/initialism can also be found in the term 'EPCM' (standing for 'Engineering, Procurement and Construction Management'), in which the key/operative letter that distinguishes this implementation approach is the 'M' (Management). However, an EPCM Project is an entirely different type of construction procurement arrangement to that employed for an EPC Project but, because of the occurrence of 'EPC' at its start, the term 'EPCM' often causes confusion within the global construction industry (as observed by Loots and Henchie).[6]

Under the EPCM arrangement, a Management Consultant (which is, more often than not, not a Contractor) will provide the management and coordination services for the whole of the Project on behalf of the Employer. Quite often, but not always, the Management Consultant may also be responsible for carrying out the actual engineering design services. The primary functions of the Management Consultant are to assist the Employer in the procurement process and arrange for others to carry out all the necessary construction work (sometimes engaging multiple Contractors), where the Management Consultant will:

(i) organise the procurement work (but not enter into agreements with, nor be responsible for paying, the Vendors and Contractors), and

(ii) supervise the materials/equipment deliveries and construction work (where the construction Contractors will be also appointed directly by and paid by the Employer, not by the Management Consultant).

The reasons sometimes given for employing an EPCM approach are that it can give the Employer greater control over as many elements of the implementation activities as the Employer is capable of handling (or wishes to have more control over). Such choices may cover the decision as to which concept design is to be followed, or which Engineering Team is to be appointed for the Front-End Engineering Design work and/or the Detailed Design work. Another area where the Employer may wish for greater control is in attempting to keep the procurement and construction costs down (principally by avoiding scope creep, which could otherwise lead to loss of time and also unnecessary

---

6 Loots P. and Henchie N. (2007). *Worlds Apart: EPC and EPCM Contracts: Risk Issues and Allocation*, Mayer Brown, p. 2.

increased costs). This greater role in the implementation process would quite obviously not be suitable for all Employers, since it would require a large contingent of people simply to handle the procurement process alone. The pressure on the Employer would be even greater if the Employer took on the added responsibility for carrying out the engineering evaluation review work or the bidding and bid evaluation work that would be necessary before orders for purchasing materials, goods, and equipment could be placed and the construction work packages awarded.

An EPCM arrangement therefore deals only with the provision of certain construction management services involved in the implementation of a construction Project, not the undertaking of any aspect of the physical work activities. However, it is perfectly possible under this arrangement for the Management Consultant to be appointed to undertake additional functions, such as producing/developing the Conceptual Design and undertaking the Basic Engineering Design work (or the Front-End Engineering Design work, dependent on the industry involved) as well as the Detailed Design work. The Management Consultant may perhaps even undertake the Detailed Design work (in circumstances where an EPC contractor will not be appointed, which would otherwise complicate the allocation of the responsibility for design and functionality risks). Additionally, the Management Consultant will usually be required to organise and supervise all the procurement work (again, where an EPC contractor is not to be appointed). The Management Consultant would not normally carry out any construction work (since that would automatically create a conflict of interest) but, instead, would usually be required to monitor the construction activities and act as the Employer's eyes and ears, in order to ensure that the facility is completed in accordance with the Employer's requirements in all respects.

The Management Consultant's responsibility for meeting the completion time requirements for a Project will also vary under an EPCM contract according to (i) the level of authority granted to the Management Consultant and (ii) the extent of the involvement of the Employer's Team in directing and supervising the other Project participants. Such a contractual arrangement therefore may or may not hold the Management Consultant responsible for timely completion of the Project, and the Conditions of Contract must be especially written for such an EPCM arrangement, since there are currently no standard forms of contract to deal with this approach to Project implementation.

The complexity of the possible EPCM arrangements is explored and explained in the PricewaterhouseCoopers (PwC) paper titled 'EPCM Contracts: Project delivery through engineering, procurement, and construction management contracts'.[7] A reading of other extant publications on the Internet will reveal widely different views as to the suitability of adopting the EPCM approach over the EPC approach. This means that it is therefore essential for the Employer to obtain competent advice about the most appropriate contractual route to take for a major Project, since the wrong choice could result in the Project being completed a lot later than the Employer requires, and

---

7 McNair D. (January 2016). *EPCM Contracts: project delivery through engineering, procurement and construction management contracts*. www.pwc.com.au/legal/assets/investing-in-infrastructure/iif-8-epcm-contracts-feb16-3.pdf (accessed 29 October 2018).

at a far higher price than expected. In regard to that, Ron Douglas[8] provides a series of charts to compare EPC and EPCM contracting options for the key aspects of a Project, and he observes:

> ... the inefficiencies of layering and bureaucracy of decision making processes in EPCM result in inefficient organisation for an extended time and higher costs result. Also, the risk remains with the client and recent performance history indicates owners are being impacted by the advice of their representatives.

Following that note, and having explained how the term 'EPCM' can cause confusion, I will go no further with the topic of the EPCM approach, since the primary subject matter of this book is about EPC Projects which, as demonstrated, most certainly are not EPCM Projects.

2.6 Employers Prefer Lump-Sum Contracts

I have had many a conversation over the past 15 years or so with senior Consulting Engineers and members of Project Management Consultancies as to why it is that lump-sum EPC Contracts are now preferred more by Employers than they were in previous years. As a result of those discussions, I have formed the opinion that the following are probably the main reasons:

1. Employers seem to have been persuaded that an EPC Project will be delivered much quicker on account of the probable time saving in the design stage. Ordinarily, under the Traditional Contracting arrangement, the Contractor will not be able to commence the construction work until the design has been fully completed by the Employer's Design Team. This line of thinking is anchored in the belief that a Contractor will be able to start the construction work much earlier under an EPC arrangement, since any yet-to-be-completed design portions will be under the Contractor's complete control.

2. Employers believe that, under the lump-sum EPC arrangement, there is far more certainty for them about the final costs of the Project than with the Traditional Contracting approach. This belief seems to hinge on the fact that the Contractor would be responsible for calculating its own quantities for bid pricing purposes, and would therefore not be able to claim extras on the grounds that the Employer-supplied bills of quantities were inadequate and had led to unavoidable under-pricing by the Contractor at the bidding stage.

3. Employers also believe that, with the right level of quality control (usually with the help of Project Management Consultants) and the imposition of appropriate performance/reliability requirements, the final quality of the completed facility and its reliability under operation would be no less for an EPC Project than if the Traditional Contracting route were to be employed.

I conducted reasonably extensive investigations into the above points to see if the Employers' suppositions held water, but my attempts at proving the efficacy (or

---

8 Ibid., Section 2.4.2.

otherwise) of the EPC approach over the Traditional Contracting route only turned up the following.

1. Construction Projects worldwide notoriously do not finish on time.[9] However, I could not determine whether or not EPC Projects finish closer to their original completion time than Projects implemented using the Traditional Contracting route, simply because I could not unearth any comparative data on the topic.

2. Likewise, I was not able to find any evidence to show whether or not Projects conducted under an EPC arrangement are less affected by increased costs for the Employer than those conducted under the Traditional Contracting approach.

3. When it comes to data showing that, in terms of quality and reliability, the completed facility resulting from EPC Projects is as good as or worse than that seen in Projects completed via the Traditional Contracting route, nothing seems to be available there either.

Since I could find no data to counter the abovementioned supposed benefits of EPC Contracts from the Employer's perspective, I have no reason to consider that the Employer's viewpoint might be wrong. In addition to the foregoing, it also seems that Employers very much like the fact that the EPC Contractor offers them a single point of contact ('one-stop shop') for everything on the Project. That is not just for the guaranteeing of the efficacy of the design and construction work but also for allocating total responsibility in the event that something goes wrong at any stage, whether it is with design, procurement, construction, or post-completion operations. A further added benefit for Employers is that many of the risks that would ordinarily fall to the Employer to bear can legitimately be transferred to the Contractor under the EPC Contract (at a price, of course), simply because the Contractor is given the entire Project implementation responsibility. That could never be the situation under the Traditional Contracting route.

Having worked on both sides of the fence (for Employers, both under and not under the PMC umbrella, and with/for Contractors), I long ago concluded that, on balance, lump-sum EPC Contracts benefit the Employer more than the Contractor. This conclusion was not based solely on the fact that I did not see any single one of the Contractors I had been associated with make the level of profit that was anticipated at the time the EPC bid was submitted (some even made large losses).[10] More than anything, my mind was made up because I had many times observed that, provided that the Employer knew with certainty what was needed in the completed facility and avoided issuing too many sizeable Variations, the final cost for an EPC Project was far more likely to come within the Employer's original budget than if the Traditional Contracting route had been adopted. This greater certainty regarding costs for the Employer is mainly due to the fact that, under the Traditional Contracting approach, the Employer has little control over the Design Team but still has to bear all the extra

9 Agarwal R., Chandrasekaran S. and Sridhar M. for McKinsey & Company (June 2016). *Imagining construction's digital future.* http://www.mckinsey.com/industries/capital-Projects-and-infrastructure/our-insights/imagining-constructions-digital-future, p. 1, Exhibit 1 (accessed 23 August 2017).
10 My conclusion does not apply to 'reimbursable' type EPC Contracts (which nearly always benefit the Contractor, but, sadly for Contractors, seem to be next to non-existent nowadays).

costs of design changes and any extras introduced by the Design Team, plus bear the extension of time and all the associated costs any such matters cause.

On the other hand, under an EPC Project, if the extension of time for completion of the Project is not caused by an excusable delay, the Contractor has to bear the associated additional cost burden as well as compensate the Employer (by way of Liquidated Damages) for completing the Project late. I can therefore fully appreciate why Employers nowadays will want to turn more and more to undertaking new complex Projects on a lump-sum EPC basis, rather than rely on the Traditional Contracting route. All this is not good news for Contractors since, as I have just mentioned, the risks placed upon them are far greater with lump-sum EPC work than where some other entity is responsible for undertaking the design work. For the majority of EPC Projects, those increased risks tend to make it harder for the Contractors to achieve the originally intended level of profit.

For anybody wishing to obtain an easy-to-read overview of the EPC and Design-Build models and the benefits thereof from a government's perspective, there was a very helpful paper presented at the Project Management Conference held at Canada's Chateau Nova Hotel, Yellowknife.[11] It provides a general discussion of the approaches and deals with the Employer's considerations, including the risks and problems involved. It also offers a range of mitigation measures that can be adopted by the Employer and/or the Contractor. In addition, it sets down a lot of questions that it would be best for the Employer and the Contractor to obtain answers to before committing to entering into an EPC or Design-Build contractual arrangement. However, overall the paper seems to be very supportive of the use of EPC and Design-Build Projects for bringing more certainty of success from the Employer's perspective. This is what helped tip the balance towards me concluding that EPC Projects favour the Employer far more than they do the Contractor.

2.7 Fixed-Price Lump-Sum Contracts

Sometimes, the Conditions of Contract proffered by the Employer for EPC Projects may contain the words 'fixed-price' as well as the term 'lump-sum'; I have seen that this combination of words can hurt the Contractor. I say this, because I experienced a number of instances where the Employer's Team took the position that those additional words meant that the Contractor could not claim any additional money over and above the original Contract Price under any circumstances whatsoever. That position ignored the fact that there were many clauses in the Contract that mentioned the right of the Contractor to be reimbursed for extras arising from the Employer's side.

On one such occasion there happened to be (deliberately) no provisions included of any sort for Variations to be made, the reason being that the Employer's Team had made it clear that there was absolutely no contingency fund and no budget allowance for Variations to occur. The Contractor had understood that this meant that the Employer would

---

11 Johannsen H., for Singleton Urquhart LLP (21 November 2017). *Design-build /EPC contracts*. www.inf .gov.nt.ca/sites/inf/files/resources/design-build_epc_contracts_-_pmc_2017_helmut.pdf (accessed 5 May 2018).

not change a single thing from what had been agreed as the work scope incorporated in the Contract. However, it soon became clear that that was not how the Employer interpreted the Contract. The Employer's view had been that the Contractor should have included an allowance in the bid pricing for all the extras that the Employer would require, since the Contract Price was described as being fixed-price, lump-sum. This led to many arguments and several impasses, the majority of which were never satisfactorily resolved from the Contractor's perspective.

My advice to the Contractor regarding these unacceptable (and added risk) situations is twofold:

1. Never accept the words 'fixed price' in the Conditions of Contract, and most certainly not in conjunction with the phrase 'lump-sum'.

2. Always insist that adequate clauses are added that will entitle the Contractor to reimbursement for Employer-directed Variations. This is especially so if the Employer is adamant that such clauses are missing because no Variations will be allowed (which should be regarded by the Contractor as a 'red flag' that major problems are sitting just over the horizon).

2.8   Selecting the EPC Contractor

In an effort to select the best Contractor for an EPC Project, the Employer will normally issue a suite of appropriate bidding (tender) documents to a group of prequalified construction-focused bidders that are highly experienced in, and can demonstrate satisfactory execution of, EPC work. Those bidding documents set out the specific requirements of the Employer for the work to be performed, as well as the requirements for the end outputs from the completed facility. The bidding process itself is frequently divided into two distinctly separate parts (the Technical Bid Proposal and the Commercial Bid Proposal), where the Employer will not proceed with considering a Commercial Bid Proposal until the corresponding Technical Bid Proposal has been found to be acceptable.

The Contractor's preparation work for its Bid Proposals will usually be undertaken under the direction and control of its Proposal Department, and it will also involve a range of different people from the company's corporate workforce (i.e. personnel who are part of the main office's permanent staff and who will not usually be allocated to work on any specific Projects). A tremendous amount of work goes into preparing both the Technical and the Commercial Bid Proposals, and it is not unusual for the EPC Contractor procurement process, from floating of the bid enquiry by the Employer to signing of the Contract, to take more than a year to complete.

The Technical Bid Proposal is compiled with input from the Contractor's technical teams, after which it is subjected to evaluation by the Employer. The main purposes of such an evaluation are:

(i) to establish the Contractor's capability for executing the Project, and

(ii) to determine if the Technical Bid Proposal meets the Employer's requirements

If the Technical Bid Proposal is found to be acceptable, the bidder will then be requested to submit the Commercial Bid Proposal. In the case of the lump-sum portion of the bid, its components will comprise priced line items for the various elements encompassed within the specified work scope. In addition, there will most probably be a requirement for the Contractor to price a schedule of rates that the Employer has provided, which would then be used later to assess the value of potential Variations. Owing to the high level of competition nowadays for new overseas construction Projects, the average industry profitability level priced in for EPC Projects is not very high (ranging from 2.5% to 7.5% on net costs, depending on the perceived level of risks involved). However, in one particular study conducted by Galonske and Weidner in a seemingly good earnings window for the plant construction industry, the actual ratio of Earnings before Interest and Taxes (EBIT) to Net Revenue earned was found to vary from +6% down to −4%.[12] This demonstrates just how tough it is for Contractors to make a profit on EPC Projects.

Moving forward, once the Employer has accepted a bidder's Technical and Commercial Bid Proposals, the Employer will generally issue a Letter of Intent to the selected Contractor before the formal Contract is signed, in order to save time in commencing the Project's implementation work. However, before the Employer issues such Letter of Intent, many negotiations and discussions will have taken place regarding the wording of the contractual and work scope clauses. In fact, a lot of effort and resources are put into that exercise, right from the stage of compiling the Bid Proposals through to the receipt of the Letter of Intent. Those costs, along with all other costs of bidding, are solely borne by each bidder at its own expense, and they are generally recorded (written off) in the financial statements of the bidders as management expenses.

It is usual for all the bidding work to be conducted in the main offices of the bidders in their home countries, even where a bidder is already working in the country where the Project is to be built. Primarily this is because it is much cheaper to do so, but also because it is easier to control the bidding activities than it is if they are conducted remotely from the main office. Even so, the bidding costs for the average Contractor are large, take up a lot of people over a long duration (as I mentioned earlier, 12 months is not unusual). Sadly though, for the typical Contractor, the bidding success rate is low; only about one in five or so bids will result in a Contract award (unless, it seems, it is a Korean company).[13]

---

12 Galonske B. and Weidner W. (2010). *Profitability of plant construction – risk management as a profit driver*. http://www.oliverwyman.com/content/dam/oliver-wyman/global/en/files/archive/2011/Risk_management_as_a_profit_driver_Perspectives_2_2010_en.pdf, p. 14 (accessed 22 October 2017).
13 Saipem (2014–2015). *General EPC contractors in oil & gas markets*. http://www.diem.ing.unibo.it/personale/saccani/index_files/Impianti%20Meccanici%20T%20(dal%202014-2015)/Il%20processo%20EPC.pdf, p. 4 (accessed 1 June 2018).

Chapter 3

EPC Project Risk Management Overview

3.1 Project Risk Management – Definition

Before proceeding further, I feel that it is pertinent to table my definition of what Project Risk Management is in regard to a Contractor undertaking a lump-sum Engineering, Procurement and Construction (EPC) Project, since it is the driver of the observations and advice that follow within this book.

Project Risk Management is the systematic process of:

(1) *proactively identifying the major potential hazards/risks for the Project as early as possible (and revisiting/redoing that exercise regularly throughout the entire duration of the Project);*

(2) *properly analysing and assessing each of the identified hazards/risks in turn, so that sound commercial decisions can be reached as to what the most appropriate response to each should be;*

(3) *putting in place adequate mitigation measures where necessary to counter those hazards/risks; and*

(4) *regularly monitoring the effectiveness of employing the hazard/risk mitigation measures and making appropriate modifications if later those measures are found to be ineffective.*

3.2 Construction Project Hazards Abound

The sheer number of hazards that Construction Projects are surrounded by is enormous, and they exist in many different forms; for overseas Projects the list of hazards is much longer than for home-based Projects. Many of those hazards simply cannot be avoided – they are just part of the territory – and so the Contractor has to develop plans for ensuring that such hazards are not allowed to transform into regrettable events. For example, where work is taking place in confined spaces (such as inside tanks) it is essential that somebody is deputed to stay close by outside and is provided with appropriate rescue means/equipment in case the worker inside gets into difficulties. The following are just a few more limited examples of where Project hazards can be found.

1. Some hazards may threaten the Project's progress, such as delays in the engineering work, delays in the procurement work for key equipment and materials, construction delays, commissioning delays, and start-up delays. Hazards having such effects include, but are certainly not limited to, the following:

   (i) an inadequate and/or inexperienced Design Team;

   (ii) a dramatic change in the economy leading to increased demand for products and consequent higher costs (requiring extra time to shop around for better prices);

   (iii) importation and transportation delays owing to bad management of the logistics support activities;

   (iv) inadequate or poor-quality labour resources (particularly in respect of local Subcontractors); and

   (v) poor quality control of construction work (leading to long lists of 'Punch Items' and the consequent late redoing of critical work).

2. In countries where the political landscape is unstable, there may be the potential for additional significant hazards:

   (i) an armed attack on the accommodation camp or the worksite, thereby putting the security of personnel at grave risk;

   (ii) importation of materials, goods, and equipment becoming a major problem, due to problems with trade tariffs or arguments over import duties payable, or even due to the lack of newly-required 'support' documentation (especially when dealing with governmental authorities);

   (iii) the local communities blockading the worksite if they feel aggrieved at the perceived high numbers of foreign workers while the locals themselves do not have jobs; and

   (iv) currency exchange rates unexpectedly and dramatically moving in the wrong direction (from the Contractor's perspective).

3. There will be everyday hazards that impact the safety of the on-Site construction workers, such as working at height, operating heavy equipment, lifting heavy loads, working in deep trenches or in confined locations, pressure testing of pipework and equipment, energising high voltage electrical equipment, etc. There could also be hazards that will impact the environment, such as spillages of toxic materials, chemical leakage into water systems, etc. These all fall under the domain of a Contractor's Health, Safety, and Environmental (HSE) Management Department, in many of which very sophisticated procedures and routines are employed nowadays. Top priority must always be given to safeguarding individuals from harm, while at the same time protecting the environment. *Thus, since these standard day-to-day risks should already be competently managed by the Contractor's designated HSE personnel, those risks are not the direct focus of this book.*

The hazard situation for construction Projects will also be subject to change over time. Items that were previously discounted as being of no significance may therefore become elevated to a critical status later. On the other hand, items viewed as being potentially dangerous to people or the Project may not materialise into actual problems. For this reason, it is essential for key personnel to be designated as part of a specialised team (the Risk Assessment Team) made responsible for continuously monitoring the changing risk situation, re-assessing the hazards as the Project progresses, and putting in place effective mechanisms for mitigating the principal perceived risks wherever possible. Brady envisages that the activities of the Risk Assessment Team will include subsequently reviewing the risks after commencement of the implementation work to aid adjusting the controls (i.e. the mitigation measures for the risks) in the light of the knowledge gained.[1]

3.3   Importance of Project Risk Management

Even as late as the First Quarter of 2019, those in the Oil and Gas Industry were still facing a tough time getting new work, primarily owing to the lower price of oil that had struck home in 2015. The impact of this fall in price was that profit margins had to be tightened dramatically in order for a bidder to stand a chance of being awarded a new Project from amongst the considerably fewer new ones coming on stream. The contrast between the abundance of mega Projects that existed in 2007 and the dearth of such Projects at the date of completing this book is very evident to those old hands still working in the Middle East since that time. The significant drop in the numbers of expatriates working there compared with years ago is felt everywhere; many of the old haunts that were previously full to overflowing are now quite desolate. However, this situation has perhaps given time for some sober reflection, since it is also widely known that the mega Projects that arose as a result of the previous high oil prices (which reached around USD 120 per barrel in 2014) were not very successful, if not outright failures.[2]

In fact, there are two very sobering key factors emerging from the latter referenced document:

(i)   78% of mega Projects failed in terms of costs and schedule, while two-thirds fell short of production-attainment goals, and

(ii)   half did not achieve at least 50% of the targeted production expected for the first 24 months of their operational lives.

Both of those factors are a clear indication that there has been a failure to manage the risks of mega Projects properly, and that Contractors must start to focus seriously on quality Project Risk Management if their thinner profits are not going to be eroded still further. It is a fact that all construction Projects have objectives at strategic, tactical, and operational levels.[3] For most lump-sum EPC Projects, the responsibility for successfully

1 Brady J. (Reprinted from *Pharmaceutical Engineering*) (January/February 2015). *Risk assessment: issues and challenges.* https://www.ispe.gr.jp/ISPE/07_public/pdf/201506_en.pdf (accessed 11 September 2018).
2 Colhoun C. and Haidar T. (for Oil & Gas iQ) – (15 April 2017). *Death Of The MegaProject?: How to Solve the Problems Involved in the Oil & Gas Industry* (accessed 26 April 2017).
3 Riskope Blog (3 April 2014). *Let's define strategic, tactical and operational planning.* https://www.riskope .com/2014/04/03/lets-define-strategic-tactical-and-operational-planning (accessed 5 September 2017).

handling all of those objectives falls to the EPC Contractor (who will also be burdened with severe penalties if any of the primary objectives are not achieved). Anything that could make achieving those objectives uncertain must therefore be considered to be a potential hazard that presents a risk, and which therefore needs to be countered as effectively as possible by the Contractor.

3.4 Corporate Risks Versus Project Risks

When dealing with risks for construction companies, there are two levels of risks to manage:

1. **The Corporate Risks**

   These are the risks related to running the day-to-day business of the Contractor and they are the direct responsibility of the Corporate Management Team (under the auspices of the Board of Directors responsible for managing the company's business). Collectively, these risks extend to taking responsibility for ensuring that all the commercial and contractual provisions included in the proposed contracts for the Projects to be taken on are sound in all respects, particularly in regard to indemnities and liabilities. Those risks are extensive, but they are also not the focus of this book, primarily because the responsibility for handling such issues should fall to specialists in the area of financial and legal matters, and not to the members of the Project Implementation Teams.

   Sadly, however, it is often management of the corporate risks where Contractors fall down. Too often, therefore, workers have unexpectedly found themselves out of a job due to the incompetence of their Corporate Management Team and/or Board of Directors, and without any financial cushion to tide them over until a new job comes along. Worse still, it is not unknown for Directors to have received bonuses (or have still been expecting them) even as they walked away from the bankruptcy that their incompetence may well have created.[4]

2. **The Project Implementation Risks**

   These are the particular risks specific to implementing each of the Projects that the Contractor is handling, and which are predominantly the responsibility of the individual Project Managers and their associated Department Managers to handle.[5] Those are the risks that this book is principally focused on.

So, instead of swamping the Department Managers with information that is more related to corporate risk management concerns, I have concentrated in this book on those risks that each of the Department Managers really does need to concern himself/herself with, especially in regard to what are known as EPC/Design-Build

---

4 Duell M., Tingle R. and Ferguson K. (For MailOnline) - (15 January 2018). *Official probe is launched into Carillion fat cat bosses who changed the rules so they could keep their huge bonuses as their company crumbled leaving 23,000 jobs at risk and taxpayers facing a billion-pound bill.* https://www.dailymail.co.uk/news/article-5269827/Government-contractor-Carillion-goes-liquidation.html (accessed 21 February 2018).

5 Fink D. (2013). *Project Risk Governance: Managing Uncertainty and Creating Organisational Value,* Chapter 3.

Projects. However, as pointed out earlier, most of the risks related to both Procurement and Construction activities are common to all Projects for construction work, not just EPC/Design-Build Projects.

3.5 Greater Risks for EPC Contractors

The scope of work and the range of responsibilities and risks placed on the Contractor will vary from one EPC Project to the next. EPC Projects of themselves are risky enough for the Contractor, even if they are of the more preferable (from the Contractor's perspective) reimbursable type contracts (where the Contractor is paid for all costs involved in constructing the Project plus an agreed lump-sum or percentage addition to cover overheads and profit). As observed by Watt, 'Since the cost of the project is reimbursable, the contractor has much less risk associated with cost increases'.[6] However, all too often nowadays, Employers are insisting that the Contract Price is lump-sum in nature and fixed in all respects (i.e. no increase is payable to the Contractor for inflation, fluctuations in currency exchange rates, etc.). This means that every extra cent spent must come out of the Contractor's own pocket, sometimes even if the Employer appears to be the direct cause of the unexpected expenditure.

Some additional items added to the list of responsibilities for the EPC Contractor can increase the magnitude of the risks almost exponentially. Examples are such things as: (i) validation of the Conceptual Design and (ii) the provision of 'single source' specialist equipment that is fundamental to the required outputs from the completed facility (such as, for example, the Gas Turbine Generators for a Power Plant Project). There are far more possibilities beyond those limited examples. This situation is aggravated even more where the Contractor is also responsible for meeting the Guaranteed Performance Outputs of single source equipment. I have even worked on a Project where, albeit in a roundabout way, the Employer tried (unsuccessfully) to make the Contractor responsible for the quality of the gas that the Employer was responsible for supplying for fuel purposes, and upon which the required outputs from the facility were highly dependent. Such is the lot of the EPC Contractor, all too often 'not a happy one'.[7]

If the Employer also insists on an unusually short time-frame for completion (with substantial Liquidated Damages payable by the Contractor if the Project is completed late), then truly effective management of the risks involved becomes an absolutely essential task for the Contractor to perform if the Contractor's expected profit is going to have any chance of being achieved. The Employer will inevitably be very strict on adherence to the completion date in the situation where the Employer is liable to pay heavy penalties to any third parties in the event of delayed completion of the Project. That will be especially so where contractual promises have been made by the Employer to other companies relying on the output date from the completed facility to be achieved.

---

6 Watt A. (14 August 2014). *Project Management*. Chapter 13, Procurement Management, Contract Types, Cost-Reimbursable Contracts. https://opentextbc.ca/projectmanagement/chapter/chapter-13-procurement-management-project-management (accessed 17 October 2018).
7 With all due deference to Gilbert & Sullivan (respectively, lyricist and composer) and their much-acclaimed comic opera, 'The Pirates of Penzance'.

To put the risks of the above scenarios into perspective, consider how much easier the task of the Contractor would be if the construction implementation work were to be set up under the following ideal risk reduction Traditional Contracting scenario (from the point of view of the Contractor):

(i) all the Engineering work is to be undertaken by others (including the Detailed Design work),

(ii) all key equipment items are to be purchased directly from the Vendors by the Employer and installed under separate Subcontracts issued by the Employer,

(iii) specialist services installations too are to be undertaken by Subcontractors who will be tied into Subcontracts issued by the Employer,

(iv) a very reasonable time-frame for completion will be allowed, and

(v) no Guaranteed Performance Outputs of any sort are required from the Contractor.

Even under the above setting of much reduced risks, a Contractor would still face many risk situations which, if they materialised, would prevent achievement of the expected profit level. Many Projects under the Traditional Contracting approach do in fact have many of those idealistic elements on board. Sadly, however, for Projects considered overall (construction work orientated), KPMG reported in March 2015 that 'Owners continue to experience project failures: 53% suffered one or more underperforming projects in the previous year. For energy and natural resources and public sector respondents the figures were 71% and 90% respectively. Only 31% of all respondents' projects came within 10% of budget in the past 3 years. Just 25% of projects came within 10% of their original deadlines in the past 3 years'.[8] It is therefore very obvious that EPC Projects require a much more robust approach to Project Risk Management if the Managers of a Project are going to be able to hold their heads high when the Final Acceptance Certificate is issued.

3.6 Principal Disaster Areas on EPC Projects

In a nutshell, the 'real and present danger' for any EPC Contractor is the actual occurrence of any one of the following four disaster scenarios on its EPC Projects, since each issue mentioned is under the Contractor's direct control:

1. **Time Disaster**

   The time for completion may be severely delayed due to poor management of the Engineering, Procurement, and/or Construction activities as well as the interfaces between them. This can then lead both to unrecoverable additional costs and the Contractor's income being seriously eroded due to the requirement to pay Liquidated Damages to the Employer to compensate for delayed completion.

---

8 KPMG International Cooperative (March 2015). *Global construction survey 2015 – climbing the curve.* https://www.assets.kpmg.com/content/dam/kpmg/pdf/2015/04/global-construction-survey-2015.pdf (accessed 20 April 2017).

2. **Commercial Disaster**

   Even if the Project is completed on time, the expenditure may significantly exceed the income due to the inadequacy of the Contractor's internal controls (not only in respect of purchases and subcontracted work but also for worker productivity and wastage levels).

3. **Quality Disaster**

   The quality of the finished work may be well below the required standard. This could then lead to litigation over arguments as to whether or not the facility is suitable for its intended purpose. In the worst case scenario, it could lead to a complete breakdown of the facility and also involve loss of life.

4. **HSE Disaster**

   An untoward health, safety, and/or environmental issue may cause a severe problem that could irreparably destroy the Contractor's reputation for a good while to come, as well as being costly to remedy (or provide compensation) for the negative effects experienced.

I have observed that far too few Contractors realise that, from the moment the Contract signing has been concluded, the Contractor has entered into the equivalent of a war zone, where *time* is the Contractor's principal enemy. The different ways in which time can be lost are many, some of which are almost indiscernible. Time moves along at a constant pace and is unrelenting in its progress, no matter how badly the Contractor needs to conserve time. The steady march of time steals meaningful progress from the Contractor in many areas of Project work, and its damage is very often added to by the Contractor's Team being too complacent or too slow to react to time being stolen. Consider, for example, the following non-exhaustive list of issues where time is more usually lost on Projects, many of whose causes (although not all) could be prevented with foresight:

   (i) late submission of Engineering Deliverables for review purposes by the Contractor's Engineering Team,

   (ii) late review of Engineering Deliverables by the Employer's Team,

   (iii) late receipt of Engineering Deliverables for Procurement and Construction purposes,

   (iv) late placement of Purchase Orders,

   (v) late handing over of the Site,

   (vi) late delivery of materials, goods, and equipment,

   (vii) late instructions for changes arising from the Employer's side,

   (viii) late mobilisation of construction equipment and manpower resources,

   (ix) slow clearance of Punch Items, and

   (x) dealing with too much reworking, often at a late stage.

3.7 Maintaining the Project Schedule

Insisting on adherence to a Project Schedule (and updating it regularly to show actual progress) has nothing at all to do with being able to blame people if certain activities are delayed. It has everything to do with ensuring that the Contractor will have the necessary backup information and data available to be able to claim an appropriate extension of time if delays occur because of the Employer's action or inaction. The importance of keeping to the Project Schedule and avoiding loss of time cannot be stressed enough. This is a particularly valid observation if the Contract includes a provision for the Employer to terminate the Contract in the event that the Project is already so late against the Project Schedule that it appears that the Liquidated Damages will be fully consumed. Time cannot and never will be the Contractor's friend (and, as indicated earlier, very often time is the Contractor's worst enemy). If, therefore, the Contractor fails to develop a worthwhile Project Schedule or fails to take all steps necessary to adhere faithfully to such Schedule, then time will be given a great opportunity to destroy all the Contractor's other efforts to complete the Project successfully. Of course, saving time can be a double-edged sword, and every care must be taken to ensure that catch-up plans and acceleration arrangements do not jeopardise worker safety or health.

Still on the importance of not losing time, there is a classic error made by many Contractors in the situation where the Employer has indicated that a major change is being considered. That is to slow down, halt or re-sequence the contractual work scope, without first having received authority or formal advice from the Employer to do so. The best way for the Contractor to proceed in all such cases is to ignore the potential change and carry on as usual until such time as the Employer issues a formally signed instruction to do otherwise. I have lost count of the number of situations I have personal knowledge of where Contractors had, in all good faith, ignored the advice to continue without letting up. Instead, they delayed the work progress, only to later find that they had been held entirely responsible for the resultant delay to the contractually required completion date. All is fair in love and war, and there is rarely (if ever) a lot of love existing between the Employer and the Contractor when it comes to delayed Project completion. It should therefore come as no surprise that even the most reputable of Employers may resort to unfairly using the threat of Liquidated Damages against the Contractor in order to extract greater benefit (or action) from the Contractor than is otherwise merited.

3.8 Departmental Interface Issues

A key point to bear in mind is that the individual Departments responsible for the EPC activities are most definitely not independent of each other, even if their respective Managers believe differently. There are probably more Managers who believe that their Departments are independent, and that their Departments are their own personal fiefdoms, than you might think possible. On the contrary, there is a lot of interaction required between those Departments, and it is often at the interfaces between them where the work processes break down. The below set of items provides just a few examples of where things can go wrong at the Departmental interfaces.

3.8.1   Rework

A great deal of the loss experienced on EPC Projects is caused by the need to redo work, and much of that reworking is caused by the lack of integrated working between the various Discipline Engineers within the Project's Engineering Team (such as Electrical Engineers, Mechanical Engineers, Civil Engineers, etc.). This can often lead to such problems as:

   (i) delays in the Engineering work due to the need to redo drawings, plus associated extra costs,

  (ii) purchasing of wrong materials, goods, and equipment,

 (iii) late ordering of the correct materials, goods, and equipment,

 (iv) taking down and disposal of wrongly installed components (and, sometimes, demolition of work already completed), and

  (v) consequent loss of time on the Project's Critical Path.

All too often, this situation arises because there is no effective way for information (documents, drawings, and data) sharing between the various Engineering Disciplines, each operating with different specialised systems that do not readily 'talk' to each other.[9] This is supported in the statement contained in the document titled 'An Introduction to ISO 15926': 'Any who have worked in an engineering environment will know that there is more than one CAD application in common use, and that on a large project all business partners do not always use the same one'.[10] This lack of system integration inevitably leads to Engineering delays that then impact horribly on both Procurement and Construction activities. As the need to reduce costs and improve efficiency in the construction industry increases, it will become increasingly important for Engineering Departments to sharpen up and ensure that lack of inter-disciplinary dialogue and communication does not contribute to the occurrence of unnecessary rework.

This will inevitably require a Contractor to invest in a good quality enterprise-wide Electronic Document Management System (EDMS) that is capable of allowing documents and drawings to be shared between all Departments in real time, so that every Project participant can be well-informed at all times. However, this can prove to be a problem for those many Contractors who do not have their own internal Engineering design capability and, instead, sublet the Engineering design work to third parties. The resultant communication gap does not allow the Contractor to obtain enough real-time knowledge at any point to have confidence about the timely delivery of the Engineering outputs. Under such circumstances, it is not surprising to find that the Engineering mistakes are very often not discovered until as late as the commissioning stage, when the problem becomes an embarrassment, as well as being more costly and time consuming to rectify.

9 Carnet E., Hultin H. and Haidar T. (for Oil & Gas iQ) – (30 October 2017). *Why your oil and gas projects are dying from rework*. https://www.oilandgasiq.com/oil-and-gas-production-and-operations/whitepapers/why-your-oil-and-gas-projects-are-dying-from (accessed 29 November 2017).
10 Fiatech (November 2011). *An introduction to ISO 15926*, p. 68. https://www.scribd.com/document/347804626/An-Introduction-to-ISO-15926-pdf (accessed 30 August 2018).

3.8.2 Delayed Technical Bid Evaluations

If the Engineering Department does not conduct the Technical Bid Evaluations (TBEs) in a timely manner, then there is a risk that the deadline for the placement of a critical Purchase Order may be missed. It would be best if the Procurement Department's personnel continually followed up on the status of all TBEs but, since most Contractors nowadays operate with minimal staffing, it is very possible that the non-arrival of an important TBE will not be spotted until it has become a highly critical issue. The Engineering Department's staff will also be under pressure to complete the engineering design work and, sometimes, the TBEs are seen as a distraction to the design activities. In addition to that, not all Engineers understand the necessity for adherence to agreed activity time-frames and, therefore, constantly need to be chased in order to get their outputs delivered on time.

3.8.3 Late Mobilisation of Procurement Team

A dedicated full-time Procurement Team may not have been set up quickly enough because the intended members of that Department are either busy on other Projects or have not yet been brought into the Contractor's company. This can happen on an EPC Project simply because the Procurement Department will not be required to be at full capacity until such time as a suitable number of Materials Requisitions (MRs) have been completed by the Engineering Department. This means that the staffing for the Procurement Department tends to arrive almost on a drip-feed basis until the full complement has been achieved. Consequently, even though the date will have been set for when the MR for the earliest of the critical Long-Lead Items (LLIs) is to be made available, the appropriate person may not be around in the Procurement Department in time to sound the alarm bell about the probable lateness of that MR. This may then become a major problem that is difficult to overcome. However, simple questioning of the Engineering Department's Work Package Engineer by the Procurement Department's Buyer would usually very easily establish whether or not the required MR will materialise on time.

I have experienced the above embarrassing situation occurring a number of times, and it was almost always impossible to make up the lost time for the LLIs involved. It was not the direct fault of the Procurement Department, since the delays/snags had occurred in the Engineering Department. However, the fact that the person next in line (the Buyer) was not around to chase for the completed engineering information meant that the Engineering Department's problem was allowed to drag on unresolved for too long. Nowadays, even if no EDMS has been set up for more direct communication purposes, a standard emailing alert system (such as Microsoft Outlook) can be utilised to issue reminder notices to check that essential activities are being progressed as required. This simple technique could prove of great benefit to the Procurement Manager before the full Procurement Team is in place on a full-time basis. However, I have yet to see even such simple, readily-available technology being utilised adequately. Ironically though, I have seen somebody who failed to use Outlook to remind them of an important work deadline employ it to remind them to leave work early so as to be on time for a doctor's appointment.

3.8.4 Red-Line Drawings Left Too Late

If the Construction Department does not complete the Red-Line Drawings in an orderly or timely fashion, then there is a very real risk that the commissioning activities will not be allowed to commence on time due to the lateness of the As-Built Drawings that the Engineering Department is required to produce. This problem is exacerbated in the situation where the Field Engineering work (particularly incorporation of the contents of the Red-Line Drawings) is not done on the Site but, instead, is conducted at the Contractor's Home Office. This problem is made even worse if that Home Office is in an overseas location compared to the location of the Site.

3.9 Forging an Integrated Implementation Team

Another area of Project working that is all too often overlooked or not accorded enough attention is building an integrated implementation team that is working to the same objectives. I have observed that if the people on the tools can see that the Project Management Team members are concerned with the welfare and well-being of all staff when working in difficult overseas situations, then it can generate a good team spirit. I am not advocating that management personnel should jump into the trenches to help out with urgent cable pulling if the labour force is understaffed. However, I once saw that done to very good effect, and it demonstrated to a lot of the management staff (including me) the benefit of breaking down the traditional barriers that often still exist today between management staff and the manual workforce. I am referring more to the need to make the lives of the workers more tolerable in tough environments where they are far from their families and loved ones.

In such cases it costs very little extra to arrange a sports day where the emphasis is on fun, or a barbecue-on-the-beach day. Holding film nights with the latest movies on show also helps boost morale, as does ensuring that televisions are provided to everybody, with access to a good selection of movies, and even Internet entertainment (with a stable connection). Providing wholesome and tasty food too, as well as worthwhile gymnasium facilities for excess calories to be worked off, also helps people feel more comfortable when they are working away from home. The list of other possibilities for making the lives of the workers better is long, and it just needs a little imagination and effort to give effect to such ideas.

The opposite approach to the above is if the workers on the Site are simply being pushed to get the Project finished on time without necessary additional resources being provided, and with little done to provide relief for the workers in their free time. This very quickly leads to workers being demotivated, and both productivity levels and work quality can then suffer as a result. In addition, and far more importantly, more risks may be taken by the workers due to their weariness. This same pressure can of course be applied to the Engineering and Procurement Teams with their office work. However, the consequences are much more likely to be the need to redo work or place 'top-up' orders for additional materials (with consequent higher cost implications), as opposed to risking the safety and health of the workers involved. Nonetheless, the additional stress that may be caused by constantly working under high pressure should not be an insignificant matter of concern. I myself have seen a number of good people fall by the wayside due to pressure of work. The safety, health, welfare, and general well-being of the workers should therefore be of the highest importance to the Contractor, and it requires a lot of effort and dedication from the management team to be successful in keeping the workers contented.

Of course, worker safety must be seen by all to be a critical aspect of how the Contractor operates. However, The Contractor must be careful that this is not viewed as purely a numbers game but, instead, it must be clearly shown that the HSE efforts stem from a genuine desire to keep the workers safe at all times. Aligning the productivity desires of the Contractor's Project Management Team with the requirement to ensure the safety of the workers means that much more attention needs to be given as to how the individual on-Site activities are performed.

It is fine to aim for finding ways to get the job done quicker, but that must be achieved by not adding risks to worker safety. In addition, working quicker must not compromise the quality of the finished product. This requires the management team to discuss detailed work procedures with the workers themselves, in order to fully explore improvement possibilities. However, I have observed that such steps are rarely taken. Unless such integrated team working is instituted from the Corporate Management Team all the way down to artisan level, a Contractor's chances of achieving top-notch 'Operational Excellence' (OE) will never materialise. In regard to OE, I particularly found Price's definition of OE very appropriate: 'building a sustainable competitive advantage through operations management' since, as he says, that objective 'accommodates process, culture, and results'.[11] Since working properly towards achieving OE promises to be an almost sure-fire way to reduce risks (not least for the manual workforce), as well as increase profitability for the Contractor, I for one consider that OE is a goal worth striving for. I am therefore hoping that the concept catches on in a big way.

3.10 Allocating Responsibility for Handling Risks

I have seen a steady movement over the past decade or so to change the perception of Project Risk Management away from it being the simple application of common sense and more to it being an intricate management system that can only be applied by consultants who are specialised in the field. Perhaps this is because, as Professor John Adams has observed, it is not only that the word 'risk' 'engenders a sense of urgency because it alludes to the probability of adverse, sometimes catastrophic, outcomes',[12] but also, as he adds: 'people are using the same word, to refer to different things, and shouting past each other'.[13] I have found that this trend has added some confusion about whose

---

11 Price R. (19 February 2015). *What is operational excellence and how is it measured?* https://www.eonsolutions.io/blog/what-is-operational-excellence-and-how-is-it-measured (accessed 2 July 2018).
12 Adams J. (For The Social Affairs Unit) (10 March 2005). *Risk management: it's not rocket science – it's much more complicated.* http://www.socialaffairsunit.org.uk/blog/archives/000318.php (accessed 11 June 2018).
13 Ibid.

responsibility risk management is within a typical construction company. Consequently, this has been detrimental to the adoption by Contractors of a cohesive approach to implementing Project Risk Management, other than in respect of where physical (manual) work is involved (such as, for example, HSE matters, commissioning activities, etc.). Nowadays (I am very pleased to see), hands-on specialists/professionals in those particular fields (not pure risk management specialists) are engaged to manage the very real and ever-present day-to-day risks involved in such activities.

In order for a Manager to be held responsible for managing all the risks involved in his/her Department's work, he/she must first understand precisely what those particular risks are. However, I have rarely seen specific training given to Managers as to where the major risks relevant to their respective Departments sit. Without boundary limits for risk management being provided, the area of risk to be managed by Managers can appear to be very wide, and much more than the average Manager would be able to handle. That is probably a major reason why a good number of the Department Managers I worked with genuinely believed that the 'upper' management levels of their corporation were responsible for taking care of all the risks involved in the business. What was not well understood, as I discovered through questioning various Managers, is that, as Martin Lipton has stated, the Board of Directors 'cannot and should not be involved in actual day-to-day risk *management*',[14] and 'Directors should instead, through their risk *oversight* role, satisfy themselves that the risk management policies and procedures designed and implemented by the company's senior executives and risk managers are consistent with the company's strategy and risk appetite ... that necessary steps are taken to foster an enterprise-wide culture that supports appropriate risk awareness ... and that ensures that risk-taking beyond the company's determined risk appetite is recognized and appropriately escalated ...'[15]

It is therefore not surprising that I also found that a common misperception of the Managers I spoke to was that all they are required to do is ensure that others under their control 'get on with doing the Project work properly', which Hayday's findings seem to corroborate.[16] That wrong view about who is responsible for managing the risks of doing the day job is probably the main reason why the interfaces between the different Departments are often not handled effectively. I am certain that, if they are honest with themselves, most Managers would acknowledge that they are aware of instances where they themselves failed to deal with some risks properly (particularly with inter-Departmental interfaces), which then resulted in avoidable (and thus personally embarrassing) delays occurring and/or rework becoming necessary.

---

14 Lipton M., Neff D.A., Brownstein A.R. et al (posted on 8 July 2015). *Risk management and the board of directors*. https://corpgov.law.harvard.edu/2015/07/28/risk-management-and-the-board-of-directors-3 (accessed 15 April 2017, emphasis in original).
15 Ibid. (emphasis in original).
16 Hayday D. (16 July 2013). *Risk management: a process that is often avoided*. www.projectsbyyonga.com/index.php/risk-management-a-process-that-is-often-avoided (accessed 27 September 2017).

Chapter 4

EPC Project Pre-Implementation Problems

4.1 Bidding Process Pitfalls

The primary purpose of this book is to concentrate on the risks associated with the implementation of EPC Projects. However, there is one significant pre-award activity that can inadvertently create very large subsequent risks for the Contractor. That is, the undertaking of the bidding process that a Contractor must go through in order to secure the work in the first place. The biggest risks arising in the bidding process are primarily as follows.

1. Underestimating the overall time-frame that will be needed to complete the implementation work. Of course, most Managers will be fully aware that it may take longer to close out the principal EPC activities (i.e. the design work, getting the materials, goods and equipment shipped, completing the construction work and finalising the commissioning activities ready for start-up of the completed facility). However, items such as the following are often not given sufficient attention, and can then cause severe delays that impact work on the Critical Path:

    (i) the preparatory documentation and proof of overall 'Site Readiness' required, all of which needs to be approved before permission to commence the on-Site activities will be granted to the Contractor;

    (ii) the time it takes to get all the necessary paperwork together that is required to enable timely Customs clearance for imported materials, goods and equipment;

    (iii) clearing Punch Items;

    (iv) getting everything prepared correctly (including documentation) so that the commissioning activities will be allowed to commence; and

    (v) closing out all the handover procedures and documentation requirements, without which the Completion Certificate for the whole Project will not be issued by the Employer.

2. Underestimating the scope of the work involved (such as the complexity and intricacies for management/administrative inputs, as well as the quantities of physical work involved). The problem for the Contractor stems from the fact that the bid price has to be a lump-sum for the required scope of work. The work is not being done under a reimbursable contract, so no extra money will be made available if the bid price proves to be too low.

3. Not appreciating the onerous nature of those administrative clauses and requirements which are not only sitting within the Conditions of Contract but which are also buried within the other Invitation to Bid documents.

4. Significantly under-pricing the work that *has* been identified, usually as a result of not appreciating the full extent of the resources (labour and construction equipment) that will be required to undertake the work.

4.2 Failure to Embrace Lessons Learnt

The construction industry is one of the most dangerous from the point of view of injuries to, and death of, workers, as observed by Jones.[1] Added to that, the financial risks are also large, and even more so with EPC Projects that are let on a lump-sum (non-reimbursable) basis. This means that no EPC Contractor can afford to be complacent when taking on an EPC Project, and that all the Contractor's Managers must be well versed in how to handle the risks for their respective Departments.

The best way for an EPC Contractor to ensure that all major problems that have arisen on past Projects (or that arise on a current Project) are not repeated in the future, is to insist that all 'Lessons Learnt' are collected for each and every Project and shared amongst the Managers as and when they happen. When the opportunity comes to bid for another EPC Project, those Lessons Learnt should then be made available to the Contractor's bidding team to consider if any items identified therein are applicable to the new Project. From my experience, if this is not done as a deliberate, specific action, Lessons Learnt are forgotten in the haste/panic to prepare and submit the bid on time. The corporate Proposals Department must therefore put the tabling of the Lessons Learnt on the checklist of activities to be performed whenever a new Bid Proposal is being dealt with or if a new Proposal Team is being formed. Remembering key Lessons Learnt only after the bid has been submitted will almost certainly be too late. If a major problem is only identified at that stage, nothing can then be done about modifying the Bid Proposal.

However, there is not much point in simply adding new findings onto a large list of old Lessons Learnt, because nobody has the time nowadays to browse such a list to see if there is anything therein relevant to their role/function. Instead, the list should be divided up into useful sections, perhaps along Departmental lines. A contents list should be added (and even an index), so that it would be easier for people to pick out items that could be meaningful to their specific areas of work.

---

1 Jones K. (2014). *Construction leads all industries in total worker deaths*. https://www.constructconnect.com/blog/construction-news/construction-leads-industries-worker-deaths (accessed 9 July 2017)

One of the ironies with Lessons Learnt I have come across is that there are some Managers who think that passing on information about what worked best should not be part of the Lessons Learnt exercise. However, I have the opposite opinion. My reasoning is simple enough: just because a particular approach worked well on one Project, it does not mean that the Project Management Team on another Project will be aware of that. Consequently, a different approach may be tried that results in a poor outcome or even dire consequences. Learning about what worked previously would thereby reduce the risk of somebody trying out something that will not work well. Sharing both the good and the bad for 'Lessons Learnt' can only be for the ultimate benefit of a Contractor's business.

4.3   Failure to Understand Contract Terms

Tucked away inside the many documents that comprise the Contract will be the clauses that dictate how the responsibilities, obligations and liabilities of the contracting parties are apportioned. Each responsibility, obligation and liability carries a corresponding risk. It is therefore incumbent on the Project Manager and all Department and Discipline Managers to read, thoroughly understand and implement the provisions of the Contract that apply to their areas of work. Ideally, such reading should take place during the bidding stage, before the Bid Proposal is submitted. That will enable any unreasonably onerous provisions to be removed before Contract signing. Failure to read, question and understand the Contract before it is signed often leads to the Contractor incurring costs later that had not been allowed for in the bid pricing.

Reading the Contract is easy; but the problem is that, unless understanding accompanies the reading, it will become impossible to apply the Contract properly. The problem with *not* ensuring that every Manager on the Project has fully understood the provisions of the Contract is that wrong assumptions may be made about what the Contract says. For example, the logistical implications of the submission sequence/timing of the Contractor's Deliverables and the time required for reviews (and re-reviews) of same is often not fully understood at the bidding stage. The consequence of not comprehending such fundamental issues correctly from the outset is that it could lead to later expensive remedial action being required to put matters right.

After Contract signing, the Contract Administration Manager (CAM) will usually be the person made responsible for trawling through the Contract (all the documents, not just the Conditions of Contract) and compiling a comprehensive Contract Summary. That document should highlight all the main obligations of the Contractor, and it should be distributed to all the Managers to read and thoroughly digest. The CAM should also be the person to turn to if help is needed in order to interpret the Contract requirements properly. To achieve that, a special meeting should be held, led by the CAM, so that all Project participants can raise their contractual questions and get worthwhile answers provided. A good way to do this is for the questions to be put forward anonymously in advance of the meeting. That gives the chance for the CAM to prepare sound answers, and also does not reveal the name of the person raising the query. Employing that technique will help to avoid unnecessarily embarrassing the person raising the question in the situation where the answer was obvious to most of the other participants.

Another problem the Contractor sometimes faces is that the Contract may not say what the Contractor wishes it to say. However, once the Contract has been signed, it is generally too late to do anything about it. That is particularly true if changing the wording of a clause would mean that the Employer's benefits/rights are diminished. The time to implement changes to proposed contractual provisions is before the Contract is signed, otherwise it will almost certainly be too late to change even unfair clauses. This is especially important if both parties had already negotiated changes to other clauses before signing the Contract. That is because, under the latter situation, a court of law will consider that both parties had been fully aware of what they were signing. This will most likely mean that the oft-quoted legal principle of 'contra proferentem' (i.e. 'interpreted against the offeror') will be held to be inapplicable. In such a situation, a claim from the Contractor of there being grossly unfair wording in the Contract would probably fail.

A further complication is that the Conditions of Contract document is not the only place where contractual provisions are contained in a Contract, and many onerous clauses lurk in such documents as 'Administration Requirements/Procedures' and, sometimes, even in the 'Scope of Work' documents. It is therefore well worth the Contractor designating a knowledgeable person to review all the proposed Contract Documents at the bidding stage to look specifically for such hidden or conflicting provisions, as well as those sitting more obviously in the Conditions of Contract document. This would then significantly reduce the risk of nasty (and expensive) surprises being discovered later, after the Contract has been signed.

4.4 Qualifications, Deviations and Exceptions List

As mentioned above, an effective way for the Contractor to reduce the Project's commercial risks is to ensure, before the bid is submitted, that the provisions outlined in the proposed Contract are vetted thoroughly to identify unacceptable and potentially catastrophic risks. Having read all the contractual, commercial, technical and administrative requirements contained in the Invitation to Bid documentation, there will inevitably then be a large number of provisions that the bidder finds impose unacceptable risks. Those risks should be compiled into a comprehensive listing of Qualifications, Deviations and Exceptions (QD&E), which should be submitted along with the Technical Bid Proposal (and upon which the Contractor's bid pricing should be based).

A problem that often arises is that the Employer's Team will insist that the entire list of QD&E items is withdrawn by the bidder, failing which the bidder's Bid Proposal will be rejected. Where the bidder has raised fair objection to unreasonable requirements from the Employer's side, the requirement to withdraw such objection is, of course, unjust. However, the Employer's Team members (especially those of a Project Management Consultant (PMC)) will consider that they are simply protecting the Employer's best interests. The way for the bidder to overcome this hurdle is therefore to ensure that all the items in the list of objections are absolutely reasonable and will not be considered either frivolous or as simply representing an unrealistic wish list. It must also be borne in mind that all the other bidders are bidding on the same contractual terms. This means that any major deviation of those terms by one bidder alone would very likely make it impossible to evaluate the bids received adequately. This is why it is a golden rule for the Employer's Team that all bids must be submitted on an equal basis. The bidder must therefore ensure that what it submits does not breach that golden rule or, otherwise, risk all its QD&E items being rejected.

One way round that latter problem may be for the bidder to carefully word the explanation for requiring each of the QD&E items to demonstrate that they would benefit not only the Contractor but the Employer too. The bidder may then be able to get the Employer's Team to agree to amend the contractual terms for the benefit of all bidders. If, after having taken pains to provide adequate justification for the QD&E items being accepted, a PMC still rejects the bidder's QD&E items in totality, the bidder should insist on meeting the Employer face-to-face. At that meeting, the bidder should then put up a sound case to demonstrate that the Employer would benefit from modifying the contractual provisions along the lines suggested by the bidder. Once QD&E items have been withdrawn by the bidder, it will thereafter be nigh on impossible to redress any of the contractual provisions that were highlighted as being unfair at the bidding stage.

It is inevitable that the Contractor will be required to fight hard to get amended provisions into the final Contract documentation. Not least, this will be because a number of the unreasonable contractual provisions benefit people in the Employer's Team at the expense of the Employer. For example, if the Employer's review requirements could likely prolong the review cycle for Deliverables to the point where the Project's completion date would inevitably be extended, then the Employer could be deprived of timely Project completion. In practice though, some members of the Employer's Team (especially those working for the PMC) would most likely be very pleased if the Project completion date could be so easily extended, since it would provide them with a longer period of secured employment. Regrettably, my cynicism in regard to this particular matter has been forged through raw experience.

An example of where the foregoing can occur is in respect of resubmitted documents being subjected to the same review period as if they were original submissions. A further example is where, following receipt of comments from the Employer's Team against the previous version, the Employer's Team is then free to make adverse comments on resubmitted documents against information/data that was already shown on the previous version of the documents. Neither of those provisions is reasonable or fair from the Contractor's standpoint. Therefore, because of the somewhat political nature of some of the QD&E items, the task of persuading the Employer's Team to accept them should ideally be undertaken by people from the bidder's side who will not be responsible for implementing the Project after Contract signing. This approach will help avoid the situation where bad blood created before the Contract is signed spills over into the Project implementation work (as I have seen can sometimes happen if a PMC is involved).

4.5 False Management Resourcing Plan

Many bids for EPC Projects require the bidders to submit a management resourcing plan/schedule to demonstrate the anticipated inputs for the people shown on the bidder's 'Organisation Chart'. This is so that the Employer can be assured that adequate numbers of people of the right calibre will be assigned by the Contractor to manage the implementation work. However, I have seen a number of Contractors submit a resource plan that is perfectly adequate (and sometimes more than adequate), but where the number of supervisory personnel is drastically reduced once the Contract has been signed. While they may get away with such tactics for less sophisticated Employers, the International Oil Companies are far more savvy, and they often put in penalty clauses (with large financial implications) if key personnel are changed/removed from the list contained in the accepted bid submission documents. The Contractor should not make the mistake of using this tactic to aid being awarded the Project, because it would undoubtedly sow the seeds of distrust amongst the members of the Employer's Team right from the outset of the Project.

Added to that, putting in large numbers of supervisory staff in the bid submission documents that will not then be provided on the Project will only serve to aid the Employer's Team when they look for reasons to reject a later claim from the Contractor for a time extension. This is because the Employer will have been given the ammunition to argue that the reduction in the Contractor's expected supervision level was the primary reason for any delay, since it directly led to lower productivity levels from the workers. Giving the Employer that opportunity will serve only to negate the Contractor's attempt to argue that it was over-zealous quality or safety checking conducted by the Employer's Team (or other Employer-side causes) that had led to any reduced productivity levels observed.

The essential thing to note here is that the anticipated management resourcing plan included in the bid submission documentation needs to be as close as possible to the actual final numbers that the Contractor intends to employ on the Project (and the costs for that faithfully included in the bid pricing). Anything deviating too far from that will inevitably lead to problems for the Contractor in the long term.

4.6   Underestimating the Costs

When it comes to compiling Bid Proposals, it is well to keep this old adage continually in mind: 'The lowest bidder is generally the one who has left out the most when compiling the costs'. In order to be in a position to reduce the chances of all the foregoing problems occurring, the Contractor must set up a panel to review the Technical and Commercial Bid Proposals in depth. The primary objective of that review panel must be to find and correct any weaknesses that exist before those Proposals are submitted. Because of the enormously negative repercussions that can occur if an under-priced Proposal is submitted, this is not something that should be treated lightly. Highly experienced personnel should therefore be brought to bear on this task, to check that all required work elements and activities have been identified and adequately priced for, with sufficient time allowed for that vetting work to be completed comfortably. Then, after all those checks have been made, consideration must be given to the amount of additional time that needs to be built into all key activities shown in the Project Schedule to cater for the Contractor's own unexpected delays. Such additional time would then help to overcome any unexpected drops in productivity, such as when there is no choice other than to reschedule an activity to take place within a period when unfavourable weather conditions are likely to occur. Likewise, adequate 'pricing contingencies' should also be added into the bid pricing to cater for the inevitable unexpected items of expenditure, or even over-expenditure, on items that were supposedly already known about.

One thing that the Contractor needs to be aware of is that, according to the level of transparency that exists in regard to the sharing of data with the Employer, the amount of additional time built into the key activities to cater for in-field production losses may be accessible by the Employer's Team. I am referring specifically to data relating to both the resourcing levels and the productivity levels that the Contractor has told the Employer about. This is because the extent of the additional time added in for any given activity could theoretically be derived by the Employer's Team. The calculation is easily done by applying the Contractor's expected productivity norms and anticipated labour/construction equipment resources against the quantity of work involved, and then comparing the output to the time allowed in the Project Schedule for completion of the activity. The Contractor therefore needs to weigh the advantages and disadvantages of being so transparent very carefully. The advantage of the Contractor being transparent about such matters is that later, if the Employer causes disruption and/or delay that then extends the duration of an activity to the point where the Project's completion time is negatively impacted, there will be far less argument about the extension of time due to the Contractor.

One area I have found that rarely has adequate funds allocated to it is the requirement for the Contractor to prepare and submit the numerous administrative reports that are required during Project implementation. The Employer's Team needs such reports to be able to complete its monitoring and checking of the Contractor's progress and also to satisfy itself regarding the Contractor's adherence to the contractually required quality, time and safety aspects of Project implementation. Another area that I have found to be subject to a shortage of funds is the requirement to set up a proprietary Information Management system that is equally accessible to the Employer's Team. Part of the reason for this is that there are still Managers today who believe that Information Management is the job of a standard Document Controller, and not the highly specialist function it actually is. If those specialists are not put in place, it usually results in a lot of overtime working being necessary at the close of the Project, simply to play catch-up in order to fulfil the data logging requirements that are essential before operations (and sometimes even commissioning of certain components) can commence.

The above will require putting direct responsibility for scrutinising the many different bidding documents (including the specialist sections therein) onto the shoulders of people who are truly capable of reviewing them properly, and then holding them responsible if, after bid submission, it is discovered that an obvious key issue was overlooked. The principle of a Corporate Management Committee alone reviewing the bid submission documentation and taking full responsibility for it only takes away the responsibility from subordinate staff members for ensuring that the Contractor's interests are fully protected. That is not to say that a Corporate Management Committee should not be set up to review the bid submission documentation; quite the opposite – it is essential. However, its primary functions should be (i) to review the reports and observations of all those responsible for analysing the Invitation to Bid documents and compiling the Bid Proposals, and (ii) to question anything that looks doubtful or carries heavy cost penalties if the bidding requirements have been misunderstood. Failure to do all of this will inevitably result in the level of profit being much reduced from that anticipated by the Contractor's Corporate Management Team.

4.7 Conceptual Design Bid Pricing Problems

There are very big risks when pricing for a lump-sum type EPC Project against a Conceptual Design only, with no Front-End Engineering Design (FEED) work having been undertaken. In any such case, I suggest that something along the following lines must be clearly stated in the Technical Bid Proposal:

(i) *The Employer's requirements shall be deemed to be only those evident from the totality of the Invitation to Bid documentation. If anything else is found to be required as a result of conducting the Conceptual Design validation work, then it shall be regarded as a Variation.*

(ii) *The Plot Plan submitted by the Contractor as part of the Technical Bid Proposal shall be considered as the agreed Plot Plan unless a different Plot Plan is mutually agreed during the negotiations prior to Contract signing. Whatever Plot Plan is finally agreed to, it must be embodied into the Contract on the strict understanding that the Contractor can immediately use it for design development purposes. In other words, the Plot Plan must become 'Rely-upon Information'. Failing that, if the Plot Plan cannot be agreed upon prior to signing the Contract, then agreement needs to be reached prior to Contract signing that the effective commencement date for the Contract is to be extended until the day following that on which the Employer's approval to the Plot Plan is given.*

I have seen very clearly that the above qualifications to the bid submission are essential in order that the Contractor will be able to recover any associated additional costs involved and also avoid the requirement to pay Liquidated Damages for delayed completion.

4.8 Agreeing to Inadequate Completion Time

Having worked for many years on the Employer's side, I am fully aware of the misguided advice sometimes given by PMCs to Employers regarding the time-frame for completion that needs to be incorporated into the Invitation to Bid documents. The misconception is that it will be of greater benefit to the Employer if the time-frame allowed for the Contractor to complete the work is made as short as possible. Occasionally, that line of thinking is coupled with the statement to the bidders that the completion date is critical, together with the stipulation that any bid submission not complying with the Project's completion time requirement (or even offering an alternative bid in respect of the completion time) will be subject to rejection. On several Projects where such thinking was evidenced in the Invitation to Bid documents, rebidding of the Project was found to be required due to the bid prices received having been far too high. Based on the fact that lower prices were obtained after the completion time requirements had been relaxed, it became very clear that the elevated original bid pricing was in order to cater for having to pay the Liquidated Damages for the inevitable delayed completion date. It probably goes without saying that the delay involved in rebidding those Projects did not benefit the Employers in any way.

One of the reasons I have heard PMCs put forward for recommending (and insisting upon) a minimalised time-frame for completion of the Project is that they sincerely believe that it does not matter how long you give a Contractor to complete a Project, because the Contractor will never meet the originally required completion date anyway. Those PMCs believed that there will always be unreasonable excuses made by the Contractor to claim an extension of time that the Employer will eventually accede to. However, if the concept of 'return on investment' for the Contractor's business was understood better by the staff of the PMCs, I am sure they would appreciate that no competent Contractor wishes its valuable resources to be tied up for one day longer than is absolutely necessary.

I have also observed that many people working for PMCs do not seem to realise that utilising resources for far longer than expected (such as supervisors, manpower and construction equipment), without guaranteed adequate compensation having been agreed in advance, will just become a drain on the Contractor's cash pool, which the Contractor will want to avoid as much as possible. Neither did those PMC staffers seem to recognise that the majority of delay problems on a Project will most likely be initiated from the Employer's side. After more than 50 years in the business I can only recollect one Project where the Employer did not introduce a significant disruptive change in requirements. To this day I remain convinced that the primary reason for that one success (and maybe the only reason) was that the Employer was in danger of losing a very lucrative concession if the Project had been delayed by even just one day. Time there had indeed been of the essence, about the only time I ever saw that apply for any Project with which I have been involved.

I have also seen too many Contractors make the mistake of believing that being reimbursed for the additional costs for any extension of time granted will be easily achieved. Such thinking is generally premised on the belief that the Employer will be reasonable, and will surely understand when it is demonstrated how the contractual time-frame stipulated in the Contract is not realistic. However, such thinking is itself unrealistic, since it ignores the following facts as to why the completion time requirement often proves to be inflexible:

   (i)   most Employers rely on their PMC (or its equivalent) to advise them as to what the reasonable completion period should be;

   (ii) the Employer may be locked into agreements with third parties based on the PMC's advice regarding the completion time, fully expecting the Project to be completed as per the contractual completion date;

   (iii) shareholders of the Employer may have been promised that same completion date for the Project; and

   (iv) the media may have been informed of the intended completion date, which can therefore be considered to have been written in stone as far as the Contractor is concerned.

The Contractor should not knowingly sign up to complete an EPC Project in a time-frame that quite obviously is unachievable. This is because, thereafter, no court of law or arbitrator would look sympathetically upon the Contractor if the Employer opts to deduct Liquidated Damages (LDs) due to the Contractor's late completion under circumstances where no significant changes to the scope of work were made by the Employer. Fortunately for most Contractors, there is very little (if any) chance of there being no significant changes on a typical EPC Project, and so there will probably always be a little room for manoeuvre by the Contractor and, hopefully, enough wriggle room to be able to get off the LDs hook eventually. However, signing up a Contract in the hope that the Employer may issue changes that will save the day for the Contractor is not a clever approach to doing business. It would give the Contractor an uncertain financial future for at least the duration of the Project, and sometimes for many years beyond that.

The correct way to tackle the issue of an unrealistic time-frame for completion would be for the Contractor to be bold at the bidding stage and challenge the completion time by proposing a more realistic/reasonable time-frame. If the Employer refuses to budge, then the Contractor should insist that an alternative bid with a revised time-frame can be submitted (although most Employers will still insist that the main [primary] bid must be for the originally requested time-frame).

One more vital point to watch out for is to see if the Contract contains a provision that the Employer can terminate the Contract if there is a chance that the LDs will be exhausted (based on the knowledge about the expected later completion date). If such a clause exists, then at some point during the bidding stage the Contractor must insist that such provision is deleted. If it is not removed, then the need to delete it should be included in the Contractor's list of QD&E submitted as part of its Technical Bid Proposal. The Contractor could also consider doing the following when the Employer thereafter still insists on a ridiculously short time-frame for completion, in an effort by the Contractor to remove the otherwise adverse financial risks inherent in being awarded the Project:

1. For the primary bid, undertake to complete the Project within the time-frame stipulated by the Employer, but include the pricing for completing the Project in the longer time-frame that the Contractor considers more realistic. Added into the bid pricing should also be the value of the LDs that would most likely be payable by the Contractor for the difference in the completion time-frame anticipated by the Contractor versus the completion time required by the Employer.

2. For the alternative bid, price for and undertake to complete the Project in the longer time-frame that the Contractor considers is more realistic. This means that the bid price would be exactly as that calculated using the approach mentioned in item 1 above, except that no allowance should be added in for LDs due to late completion.

Under the above scenario, it would be reasonably obvious to the Employer's Team that the potential LDs have been added to the bid price for the primary bid. However, the presence of the alternative bid will allow the Employer to conduct a cost–benefit analysis to determine whether or not the Contractor's alternative bid is more beneficial and therefore worth accepting. There will be occasions where the Employer will perhaps conclude that early completion of the Project is not as beneficial as previously thought. Of course, the one big caveat to the Contractor adopting this approach is that the bid price for completing within the originally required time-frame may not be the lowest. The deciding factor for the Contractor's approach will therefore depend on how desperate the Contractor is to secure the Project. With a few notable exceptions where rebidding became necessary owing to an unrealistic completion time being set, I observed that there was at least one Contractor who was determined to secure the Project despite the almost-impossible completion date set by the Employer's side; I imagine that same scenario will play out many times over in the future.

The question arises as to what the Contractor should do if the Employer remains adamant that the shorter time-frame must be maintained, will not accept an alternative bid with a longer time-frame, and insists upon being able to terminate the Contract if the LDs look set to be exhausted based on the Contractor's progress compared with the contractually required completion date. I have worked on the Employer's side where this situation existed on several Projects, but where too many bidders withdrew from the bid, citing the short time-frame as the reason. This left the Employer with no choice other than to amend the time-frame requirement and go out to re-bid, causing substantial time to be lost (and thereby defeating the whole purpose of insisting on a short time-frame in the first place). However, most Contractors will baulk at withdrawing from a bid, being worried that future bidding opportunities with that same Employer may be lost because of the refusal to bid. If, under these circumstances, the Contractor still proceeds to bid for the shorter time-frame and is awarded the Project, then it means that the Contractor's primary risk to completing the Project successfully has been very clearly identified. The Contractor will then have to pull out all the stops to complete the Project on time, as well as ensure that every possible reason for being granted an extension of time is pursued to the hilt.

One final word of warning needs to be issued here regarding keeping to the originally agreed completion time. This relates to believing that nobody will object to the Contractor belatedly deciding to implement overtime working to catch up lost time (even if the costs for possibly doing so had already been built into the Contractor's bid submission pricing). It will not be that simple, for the following reasons:

1. The Employer's Team and/or the PMC will most likely require being present while the Contractor is working, but they will not have allowed in their costings/budgets for paying overtime to their supervisory staff.

2. Additional visas may be required by the Contractor to bring in more overseas workers as well as additional supervisors. However, there will most likely be great objections made by the Ministries of both Foreign Affairs and Immigration to such a proposal if the required ratio of local labour to foreign labour becomes unbalanced, not to mention that visa processing will inevitably become much slower than expected under such circumstances.

3. Accommodation for additional imported workers/supervisors may be difficult to arrange at short notice, particularly for offshore installation work (which may require the provision of additional accommodation vessels).

One way around the above problem is to keep silent about it until the actual bid submission stage (unless the Invitation to Bid documents already preclude overtime working), but ensure that all additional costs for implementing overtime working are included in the bid price. However, the intention to implement overtime working must be included prominently inside the Technical Bid Proposal and be unequivocally stated. If this is done, then perhaps the Contractor may decide that there is no need to add in the value of the potential LDs to the bid price, thus making the Contractor's bid price much more competitive. Even if the Employer's Team unearths this ploy, it could still give the Contractor a good opportunity to negotiate a more realistic completion time-frame, especially if the Contractor's bid price is the lowest.

4.9 Reliance on Employer's Information/Data

Sometimes the Contractor is provided with critical technical information/data by the Employer for bidding purposes that relate to the Site conditions (e.g. borehole data, subsea surveys, etc.). In such cases, the Contractor must ensure that a clause is present in the Conditions of Contract that permits recovery of additional costs and/or an extension of time to be granted if the Contractor's costs increase and/or the Contractor is delayed due to the land or subsea conditions being significantly different from those represented by the Employer-provided documents. However, I have been involved on Projects where the borehole data was offered to the bidders 'for information only', and the bidders were required to include in their pricing for any later problems found with that data, even though access to the Site prior to award of the Contract was not permitted owing to security concerns.

Unfortunately, on more than one of those Projects, the borehole data was subsequently determined to have been completely inadequate. As a consequence, the Contractor was faced with significant additional costs to deal with the extra work involved. If the Contractor had insisted that it was only reasonable for that information/data to be treated as 'Rely-upon Information', then those additional costs could have been recovered under a Variation. Where the content of critical data cannot be checked before bidding but the Contractor must nonetheless take full responsibility for its correctness, signing a Contract would have to be considered as an act of faith/hope undertaken solely at the Contractor's own peril.

4.10 Late Approval of CDVR

If there is a requirement for the Contractor to endorse/accept the Conceptual Design by way of submitting a 'Conceptual Design Validation Report' (CDVR) or something similar, a big delay can occur between the submission of the CDVR and the date approval is given to the Contractor to commence the FEED work following review of the CDVR by the Employer's Team. This problem will be exacerbated if it is found that extras need to be added in as a result of the Contractor's findings mentioned in the CDVR. The Contractor therefore needs to state clearly in its Technical Bid Proposal that, for example, only 45 days have been allowed for the Employer to (i) review the CDVR, (ii) agree any

Variations to the Contract, and (iii) give approval for the Contractor to commence the FEED work, failing which the Contractor shall be entitled to an extension of the time for completion, with full costs thereof payable by the Employer.

Another important thing to remember here is that the Project Schedule must allow for that minimum 45 days' stand-down time for the bulk of the Engineering Team while the Employer's Team is conducting its review of the CDVR. This is because the Employer's Team may have opinions about the validity of the CDVR that differ significantly from the Contractor's views. The Employer's Team may thus refuse to accept any submissions of the Deliverables due from the Contractor until all Conceptual Design issues have been resolved in full. If, therefore, the Contractor makes the mistake of believing that productive work can be carried out by its full Engineering Team in those 45 days of waiting for approval, then there is the very real risk that a great deal of time and money will be lost by the Contractor. Some significant reworking of the drawings may also be required, as happened on one of the Projects I was involved with.

4.11 Gateway Between FEED and Detailed Design

Where the Contractor is required to undertake the FEED, it also needs to be stated very clearly in the Contractor's Technical Bid Proposal that there should be no gateway between the FEED and the Detailed Design work. This would mean that the Contractor would then be free to move straight into the Detailed Design work as soon as the FEED for each element has been approved, all without any prior approval being required from the Employer's Team or without having to wait for the full FEED work for every element to be completed first. If this concession is not given, it will inevitably lead to certain members of the Contractor's Engineering Team being required to demobilise (for a long period in some cases). However, those same people may not be available when remobilisation is finally required, leading to new staff members being added, as well as a long lead-in process ensuing (while they become familiarised with the work of the long-since departed former team members).

4.12 Extended Review Period for Deliverables

Timely review of the Contractor's Deliverables by the Employer's Team is essential if the Contractor is going to have any chance of keeping to the Project Schedule, especially if the Contractor is responsible for preparation of the FEED. This is because it will be impossible for the associated Procurement work to commence until the relevant Detailed Design work has been completed for any given element of work. If the Procurement work is delayed, then it will result in the affected Construction work also being delayed. This is probably the most significant item that could lead to critical delay occurring on a Project. Consequently, if the Contractor has not addressed this issue thoroughly before signing the Contract, then it will inevitably result in disagreement between the Contractor and the Employer's Team throughout the whole of the Engineering design phase. Even where the Contractor had prepared the FEED and had it approved, I have witnessed situations where review of the Detailed Design work still suffered from similar problems of Employer-side delay in handling the review work.

The principal reviewing problems that arise are as follows:

1. The Employer's Team is often unrealistic in its expectations as to the level of detail that drawings should contain (especially at the FEED stage), and it does not properly appreciate the significance of approval delays on the programmed time for the Engineering work. The causes here fall into the following distinctly different areas.

    (i) The Employer's Team does not differentiate between those submittals that are essential for the Employer's Team to review and those which do not need to be reviewed at all. Consequently, unless the matter is addressed prior to Contract signing, there is a possibility that the Employer's Team will regard all the Contractor's submittals as being Deliverables (and thus each document and drawing being subject to the full contractual review period). This is patently unreasonable and, as I have seen many times, will make it nigh on impossible for the Contractor to keep to the Project Schedule.

    (ii) The Employer's Team takes the review period stated in the Contract as its right to at least that length of time to review each and every submittal and re-submittal of a Deliverable, ignoring the fact that simple matters should take far less time to review than complex matters. Again, this situation is totally unacceptable. Even on the simplest of Projects it could easily be demonstrated that, if every submittal was subjected to the longest possible review period, the Engineering work would be delayed by a significant amount. That would be true even if no re-reviews were necessary (i.e. all submittals are approved against the first submission).

    (iii) Despite the inherent unreasonableness involved, the Employer's Team unrealistically assumes that the Contractor has somehow made full allowance in its programme for the Employer's Team to:

      – re-review every Contractor's Deliverable each time it is re-submitted, and

      – add completely new comments against re-submitted Contractor's Deliverables that were missed against previous submittals.

    (iv) The Employer's Team continually denies approval and adds comments that amount to 'preferential engineering' or 'wish list' items that the Employer is not entitled to, but which still take time to resolve before approval of the Deliverable is finally given.

2. The Employer's Team very often does not have adequate numbers of personnel available to deal with reviewing the Deliverables as quickly as is needed, but it will inevitably wish to hide this fact from the Contractor (who otherwise would then have grounds for claiming an appropriate corresponding extension of time in which to complete the Project). If sub-items (ii) and (iii) in item 1 above are a regular occurrence, then it is highly indicative of an under-staffed Employer's Team.

4.13 Objecting to Impractical Review Process

Following on from the above, the Contractor's Technical Bid Proposal therefore needs to spell the following out clearly to the Employer if the Invitation to Bid documents have

1. The only Contractor submittals that require to be reviewed by the Employer's Team (and which are thus classified as 'Deliverables') are those identified in the list attached to the Contractor's Technical Bid Proposal. All other submissions made by the Contractor shall be deemed to be for information purposes only and will not require review by or approval from the Employer's Team. *(The Contractor should of course be realistic in its compilation of such a list [which should be generic in nature] and ensure that all relevant documents have been included, otherwise this requirement will almost certainly meet with outright rejection.)*

2. The Employer's senior review team members should be present throughout the Engineering design phase in the same city as the Contractor's Engineering Team, so that regular contact can be made between the Engineering personnel from both sides. Further, the members of the Employer's Team should have the authority to make engineering decisions without the need to refer back to others located elsewhere (especially not overseas). *(Better still, if at all possible, the Employer's Team should be temporarily housed in the Contractor's offices.)*

3. The period stated in the terms and conditions of the Contract in which the Employer's Team has to respond to the Contractor's submittals that require approval from the Employer's Team (the Deliverables) shall be regarded as the absolute maximum. Further, the Employer's Team shall be obliged to use its best endeavours to complete all its reviews in the shortest possible time, consistent with the complexity or simplicity of the documentation to be reviewed. *(Ideally, the actual review period to apply for each Deliverable should be agreed between the Contractor and the Employer prior to Contract signing, but this is more often than not impossible to achieve.)*

4. The Employer's Team shall be allowed *only one review* of a Deliverable (provided always, of course, that it can validly be deemed satisfactory for review purposes). The Contractor shall, after receiving the review comments, be entitled to proceed with follow-on work without having to await the approval of the Employer's Team, on the understanding that the Contractor will faithfully incorporate those comments into the next revision of the relevant drawing/document or has raised valid objection against such comments. Where the Contractor has faithfully incorporated the comments of the Employer's Team into the next revision of a drawing/document and not made other revisions, the Employer's Team shall not be entitled to make any additional (new) comments whatsoever. Approval of the Employer's Team to the revised drawing/document submitted under such circumstances shall therefore be a 'rubber-stamp' process only. *(This is perhaps the most vital issue to deal with, since allowing the Employer's Team to re-review each resubmission as if it were a new document is where the Contractor is likely to lose a great deal of unrecoverable time that was never allowed for in the Project Schedule. Too many Project Management Consultants see this as not being their problem, despite the fact that the application of the principle of a resubmission being treated as a new submission can be very harmful to both the Contractor and the Employer, since it will inevitably result in considerable unnecessary delay for the Project.)*

5. If a revised drawing/document submitted by the Contractor contains revisions beyond incorporation of the comments of the Employer's Team, any new comments received from the Employer's Team may only be in respect of such revisions,

6. Adherence to the review period turnaround time is absolutely essential for the Contractor to complete the Engineering work in the time-frame allowed for in the Project Schedule. If any review delay from the Employer's side occurs to drawings/documents for items of work that sit on the Project's Critical Path (or will then sit on the Critical Path due to the Float having been eaten up), then the Contractor shall be entitled to an extension of the time for completion of the Project for each such item of delay.

7. In the situation where the Contractor considers that the comments received from the Employer's Team against a Deliverable represent 'preferential engineering' or 'wish list' items, the Contractor shall raise a 'Comments Clarification Request' detailing its objections. Thereafter, the Employer's Team shall be obliged to take all reasonable steps to resolve the issue within a maximum period of four calendar days. If resolution of the issue extends beyond that period and subsequently the comment is withdrawn by the Employer's Team, then the Contractor shall be entitled to an extension of the time for completion if the delay impacts an item that is identified as sitting on the Project's Critical Path or as having had its completion time extended beyond the available Float for that item.

If the Contractor decides for any reason not to adopt the above suggestions for simplifying and shortening the Deliverables review process, it can be reasonably expected that the Employer's Team will take far longer to complete its review of the Deliverables than the Contractor had allowed for in the Project Schedule. Under those circumstances, I know from personal experience that it would not be possible to complete the Project by the contractually required date.

4.14 Underestimating Equipment Procurement Packages

The problems for the Procurement Department can very often begin as far back as the bidding stage, long before the Procurement Manager for the Project has been selected. While it is perhaps not difficult to be able to asses all the different types of goods required under the 'bulk purchases' category, it is much harder to gauge the different types of equipment required. This is especially true if the design is at the conceptual stage only, and the bidders have not been given enough time to develop the FEED (as is often the case). The procurement package for each different item of equipment will require a Materials Requisition to be prepared (and possibly a specification, as well as Data Sheets), added to which Technical/Commercial Bid Evaluations will also need to be conducted by the Contractor. The last thing the Procurement Manager needs to discover is that the number of different equipment procurement packages anticipated for bid purposes was inadequately assessed.

I have been on a Project where the number of different equipment packages anticipated at the bidding stage was only 50 or so, but the finally required number of packages was double that figure. This put the Engineering Team under tremendous pressure, since its staff numbers were too small to cope with the unexpected increase in the workload. It is therefore important that an equipment expert is on board at the Technical Bid Proposal stage who is familiar with the type of Project being bid for. This would then go a long way towards getting the number of work packages established properly and avoiding the knock-on delay problems I witnessed.

4.15 Rejection of Country of Origin

Again, from working for many years on the Employer's side, I am fully aware of the misguided attitude of some PMCs still, today, when it comes to the purchasing of equipment and goods from countries other than Western Europe, the United States, Japan, Singapore and South Korea. This is the belief that the same product from any other location must be inferior and therefore unacceptable. That attitude may have been reasonable 30 or so years ago, in the days before Quality Assurance made big headway with the publication in 1987 of ISO 9000.[2] However, the former poor quality situation no longer applies to anywhere near the same extent today, since many reputable European and American companies have set up first-class manufacturing facilities in places such as China, Malaysia, Thailand, Vietnam, etc., operating under strict quality controls.

That attitude of looking down on the ability of other countries to produce quality goods was highlighted at a meeting I attended a few years back, where a spokesperson for the Employer asked what assurance there would be that goods procured from China would not be inferior to the same goods purchased in Europe or the United States. My colleague first pointed out that the Chinese manufacturing companies proposed were all QA certified subsidiaries of brand-name European companies. He then remarked that he had taken his family on holiday to the USA that very year, and that, in a major USA store, his son had asked for a pair of running shoes of a highly respectable USA brand that had taken his fancy. My colleague saw the price was fine and then noticed that the country of manufacture was China. My colleague then expressed his opinion to the Employer's Team that, if Chinese goods manufactured to approved quality standards were good enough for people in the USA, then they are certainly good enough for the company he worked for and ought to be acceptable to the Employer too. There were no more questions or doubts raised about the quality standards of the Chinese manufactured goods our company had proposed, and that same Employer's top brass later declared that the completed facility was of world-class standard.

In order to ensure that there will be no arguments later (after Contract signing) about what the country of origin for certain goods must be, the Contractor should supply with the bid submission a full listing of all the equipment and goods priced for, and show the proposed countries from which the goods will most likely be purchased. There should be no commitment made by the Contractor to purchase from any particular country (even if pressure is brought to bear at the negotiating table), because economic circumstances can change dramatically over the course of a year, especially currency exchange rates. The Contractor should leave its options open and then be free to choose the Vendor quotation that it considers offers the best all-round fit, taking into account the technical aspects, the delivery dates and the pricing.

---

2 Published by the International Organization for Standardization, who state that 'The phrase "ISO 9000 family" or "ISO 9000 series" refers to a group of quality management standards which are process standards (not product standards).'

If the Contractor is forced into making an early commitment to undertake hefty purchases from a specific country, there is a big risk that the Contractor will lose a lot of money later down the road. However, if it is perceived at the bid stage that significant equipment or goods will almost certainly need to be purchased from specific overseas countries, the foreign currency exchange rate issue can be overcome to a certain extent by the Contractor submitting a multi-currency bid (requiring payment from the Employer in more than one currency). Some Contractors go as far as forward-purchasing currencies just to be sure that there will be no such exchange rate problem; but the pros and cons of doing so have to be weighed carefully.

4.16 Responsibility for Governmental Problems

Many Employers opting for their facilities to be built under EPC Contracts want to be able to pass the risk of problems arising from obstinate/belligerent overseas governmental authorities on to the Contractor. This is despite the fact that it is the Employer who derives the benefit from the completed facilities, not the Contractor. Where the Contractor is a foreign entity in the location where the facility is to be built, it will generally be powerless against a difficult governmental authority. By contrast, the Employer should have a lot of goodwill and therefore be able to lobby for better cooperation from those bodies, not least because the completed facility will bring job opportunities to the local community.

The Invitation to Bid documents may contain provisions that severely limit the Contractor's rights to be granted an extension of time for completion (and associated reimbursement) in situations where delay and disruption is caused to the Contractor by governmental authorities, regardless of the fact that the Contractor may not be at fault. In such a case, the Contractor must strongly object to such provisions before Contract signing or, otherwise, risk incurring large additional costs and suffering delay for which there will be no redress once the Project gets under way.

4.17 Performance Bond Early Submission

I have lost count of the number of times I have seen the requirement in Invitation to Bid documents for the Contractor to submit a Performance Bond for an overseas Project within a period not exceeding 15 days from the Contract signing date. It seems to me that the drafters of such provisions have never had to obtain an overseas Performance Bond for their own respective companies, otherwise they would have known that even obtaining a Bond within 30 days (not just 15 days) is only possible if:

 (i) the Beneficiary and Guarantor are both based in the same locality,

 (ii) the Guarantor Bank is in a location where the Banks are geared up to issuing such Bonds on a regular basis without regulatory interference from a Central Bank or similar authority, and

 (iii) there will be no argument over the wording of the Bond.

However, a more realistic time-frame would be 45 days (not 15 days) to obtain the Performance Bond because of the following obstacles usually faced for large-scale EPC Projects:

   (i) the Guarantor Bank and the Beneficiary are in different countries, or

  (ii) the Guarantor Bank is located in a country where governmental permissions may be needed to issue the Bond, or

 (iii) the Employer's proposed wording is unacceptable to the Guarantor Bank and/or the approving governmental body in an overseas location.

An added complication to the above arises where the contractual provisions state that the Employer has the right to terminate the Contract if the Performance Bond is not submitted within the time-frame stipulated in the Contract. Frankly, I consider such provisions to be the result of skewed thinking, since the whole purpose of the Performance Bond is to act as an incentive to the Contractor to complete the job on time and to the right standard/quality in accordance with the Employer's requirements. So how can unnecessarily terminating the Contract early possibly be in the Employer's interests, especially when there is a perfectly sensible and much easier way to deal with any delay in the submission of the Performance Bond? There is also no need to tie a prohibition on entering the Site to the non-submission of the Performance Bond, since that will also simply delay the Project and thereby not benefit the Employer in any way. I saw this prohibition imposed quite recently in a case where the Employer wanted to change the wording of the Bond at the last minute. The ensuing delay was most definitely not the fault of the Contractor, but the Contractor was the entity made to suffer.

The way around the problem of a possibly late Performance Bond is for the Employer to insist that no payments will be made to the Contractor until a valid Performance Bond has been submitted, or until the amount of money withheld by the Employer against interim payments due equals the required value of the Performance Bond. The benefit of this approach to the Employer is that it would act as an incentive for the Contractor to get the Performance Bond submitted to the Employer as quickly as possible. The risk to the Contractor of not insisting on this 'retention' approach is that the Contract might well be terminated before it has properly begun.

4.18   Requirement for On-Demand Bonds

Most Employers nowadays seem to require the Performance Bond to be of the On-Demand type, where the Bond can be called upon at any time for any reason, and where the Contractor is usually powerless to prevent the Guarantor Bank from paying out. While it may be perfectly reasonable for an Employer to insist upon such a Bond in the situation where the Contractor has received an advance payment, there is nothing at all reasonable about a reputable Contractor being required to provide an On-Demand Performance Bond. My advice is for the Contractor to raise an objection at the bidding stage to being required to provide such a Bond and, instead, propose wording that is fair and more in line with that found in the FIDIC[3] suite of contracts (i.e. the Employer is obliged to give valid reasons for calling upon the Performance Bond). My reasoning for this is straightforward: the Employer is already in a position of strength, being the holder of the purse strings. Further, I have seen too many Employers unfairly use an

---

3 See, for example, 'Conditions of Contract for EPC Turnkey Projects' (Second Edition, 2017); published by the Fédération Internationale des Ingénieurs-Conseils (FIDIC).

unwarranted threat of calling upon the Performance Bond in order to obtain benefits that were not legitimate or deserved.

Another condition that some Employers wish to see included in the Performance Bond is for the Employer to be allowed to transfer the benefit of the Bond to any other entity of its choosing. Such a provision in an On-Demand Bond is not fair to the Contractor and so should likewise be objected to. Failing to get such a provision deleted could easily see the Contractor fighting an unknown third party to recover money that was called upon maliciously or fraudulently. While such matters relating to the Performance Bond wording may be considered trivial when the Contractor is trying to win a Project, they can become a source of great grief later in the day, particularly if the relationship with the Employer becomes strained. It is therefore better to get the obligations, liabilities and contractual wording of the Performance Bond right before the Contract is signed.

4.19 Import Duty Responsibilities

Sometimes, where important governmental development Projects are involved, the Employer or the Project itself may be granted exemption from payment of import duty on materials, goods and equipment (which in some countries may be referred to as 'levy exemption'). In this situation, it is imperative that the contractual wording in the Invitation to Bid documents is thoroughly checked to ensure that the same concept is evidenced everywhere that import duty is mentioned. However, it is not unknown for the Employer's standard documentation to make no mention of such an arrangement and, on the contrary, still contain wording that makes the Contractor appear to be responsible for payment of the import duty (and, accordingly, to have allowed for such situation in the bid pricing). If the levy exemption is suspended or permanently withdrawn for any reason (as may happen for political reasons), then the Contractor needs to be sure that:

(i) payment of all import duty will become the immediate responsibility of the Employer;

(ii) any delays thereby caused to the Project Schedule will be reason for the Contractor to be granted an extension of the time for completion; and

(iii) the Contractor will be fully reimbursed for all additional costs involved including, but not limited to, those associated with any extension of time as well as any demurrage charges that become payable at the port of entry.

One reason sometimes given for the Employer not wishing to put appropriate wording into the Contract to cover this situation is that the levy exemption is specific only to the Employer and cannot be passed onto the Contractor. While this may well be true, if the Contract Documents do not deal adequately with items (i)–(iii) above the Contractor may find itself being responsible for payment of the import duty without any reimbursement being due, and also being held responsible for any delays caused by suspension or withdrawal of the levy exemption.

The foregoing point is especially relevant in the situation where a different Employer's Team from the one that handled the Project bidding/award work is appointed for the on-Site implementation phase. In such a case there may well be no recognition given to the fact that the Contractor had been given verbal reassurances by the Employer during the bid negotiations stage that such levy exemption would always remain the direct responsibility of the Employer even if, later, levy exemption was suspended or withdrawn. I experienced the situation where a completely new Employer's Team was appointed mid-way through the Construction phase, where the new team patently was not happy to endorse much of what the previous team had agreed to, and where the Contractor then faced exactly the problems I have just mentioned. The value of the import duty that could be levied against a Project will always be considerable, so it is vital that great attention is paid to ensuring that allocation of the responsibility for payment of such import duty is clearly expressed in the Contract Documents.

4.20   Local Content Obligations Downplayed

Many developing countries have the problem of there being inadequate employment opportunities for their own people. Added to that, the local population has generally not been educated or trained to the level required for taking part in the construction work that is so vital to the development of the country, other than for simple manual labouring work. Nonetheless, when foreign Contractors are appointed to undertake the much needed construction Projects so necessary for the future well-being and development of such countries, they are often faced with obstruction from the local population in the vicinity of the proposed new facilities. This is generally due to the high unemployment levels that exist there. Sometimes the confrontations that take place turn violent, largely because the local population is angry that foreigners appear to be taking jobs that ought to be given to the locals (as the locals see it).

In the past, it was reasonably easy to persuade local officials that the necessary skilled personnel were not locally available, which therefore required larger numbers of expatriate personnel to be brought in. However, due to the spread of mobile telephones even amongst poor communities, local uprisings against foreign Contractors are more prevalent nowadays. To overcome the problem of disruption from the local population, many governments therefore now impose much tougher rules regarding how many foreign workers the Contractor can bring into the country for working on the Project, and how many from the local population must be employed. Often, the rules will even stipulate that the Subcontractors must come from areas in the vicinity where the facility is to be built and not from areas beyond that.

Some Contractors used to overcome the 'local population to foreign workers' ratio problem by paying for large hordes of unskilled workers to stand by all day and do next to nothing. This was considered as being the price to pay for the privilege of being awarded the Project. Of course, such a practice did nothing to enhance the skills of the local population that would help them get better jobs in the future. Today, the governments of many developing countries are all too well aware of that practice and have introduced rules to try to eradicate its use. More often than not nowadays, the Contractor is therefore required to directly employ local personnel at all the different levels inside the Contractor's organisation, including the management levels, plus institute training schemes for the manual workers for teaching them new skills (such as electrical wiring, pipe fitting, welding, etc.). All of this adds a lot of additional costs for the Contractor, as well as it taking up time and management resources.

It would be a big mistake for Contractors not to take the Local Content requirements much more seriously nowadays than they did in the past. The scene has changed dramatically over the past few years, following the drop in the price of oil, since that has negatively affected the income for those developing countries where oil exports represented their largest source of revenue. I am also certain that the Local Content requirements will not be so easily satisfied in the future as they once could be, even if the price of oil gets back to its previous highs. The local populations have seen how effective their protests for better recognition and fairer treatment have been, and there will be no turning back for them.

At first glance, the strange thing about Local Content requirements is that most Invitation to Bid documents do not stipulate any sanctions against the Contractor if those requirements are not complied with. However, due to the way it generally works, there is no need for the Employer to take any direct action against the Contractor to ensure compliance. This is because the various governmental ministries involved in issuing the visas and work permits for the Contractor's expatriate staff will ensure that the Contractor complies with the Local Content requirements or it will suffer. That suffering comes in the form of severe delays in the approval/issuance process of the visas and work permits until such time as the Contractor has satisfactorily demonstrated that all such requirements have been fulfilled. If those delays are due to the Contractor's non-compliance with the Local Content requirements, the Employer will not be obliged to grant the Contractor an extension of the time for completion of the Project. Worse still, Liquated Damages may have to be paid by the Contractor. Adequate financial allowance must therefore be built into the Contractor's bid pricing to cater for complying with the Local Content requirements.

4.21 Contractor's Bid Modifications Ignored

Very often the Employer's Team tries its best to exclude the Contractor's bid submission documents from becoming part of the final contractual set of documents (the Contract Documents). Even where the Employer's Team does agree to incorporate any of the Contractor's bid submission documents into the Contract, the Contractor's data is generally given the lowest level of priority/precedence. This often means that all the Contractor's attempts to improve on the original contractual provisions had no effect whatsoever.

It is therefore essential that the Contractor's Legal Team (which may comprise staff dedicated to the Project, in-house corporate lawyers and also third-party lawyers) fights hard at the negotiation stage to get all the key Qualifications, Deviations and Exceptions incorporated into the Contract Documents. The level of priority accorded those items must rank above that of the 'terms and conditions' of Contract in respect of precedence. Failure to do that would not only negate the Contractor's efforts to get a fair deal but could also cause the Contractor a lot of unnecessary grief during Project implementation, especially if there is a lot of money involved. It should be particularly noted that onerous terms and conditions may be sitting in documents other than the Conditions of Contract, such as the 'Scope of Work' or 'Administration Procedures/Requirements', etc.

I have often seen what I call 'the lazy approach' taken to incorporating the Contractor's relevant bid submission documents, and that is where all the bid correspondence is added into the Contract as one huge package of information. That approach only serves to complicate administration of the Project and should be avoided at all costs. The Contractor should therefore insist that all matters that materially affect the interpretation of the Contract should be worked into the wording in the appropriate section of the Employers' original bid documentation. This then allows what needs to be taken into account in the implementation stage to be quickly found/seen and readily comprehended. Doing that may add a few extra days to the preparation of the Contract Documents before they can be ready for signing, but my experience is that it is time well spent, which will pay for itself once the Project has got under way.

4.22   Relying on Carrots

Anybody who has been in the construction industry long enough will have faced the situation where the bid negotiations threatened to stall because of certain sticking points that seemed intractable. An example would be where the Contractor is pressured to accept, late in the day, an unusually short time-frame for completing the work, but where the Employer will not agree to pay the extra costs requested by the Contractor. In regard to this specific point, I have many times heard PMCs argue that reducing the time-frame for completion will always save the Contractor money, and Employers often tend to buy into this fallacy. However, such an opinion ignores the realities. First, there is an optimum number of personnel that can be used on any one Site; squeezing more people on Site causes congestion that reduces productivity. Second, more personnel will require more management staff and more local accommodation to be provided (in a situation where the off-Site area may well be limited). Third, excessive overtime working also causes reduced productivity and can lead to more accidents occurring due to fatigue of the workers. And, fourth, the Contractor needs a meaningful level of contingency funds available to cover the very realistic likelihood that Liquidated Damages could still become payable.

As a means of overcoming such impasses, the Employer sometimes proposes a future opportunity that sounds like a very good proposition. The Contractor may accept such a proposal in good faith, even though the offer is only verbal, not written. However, because such an offer will also not be written into the Contract, it must be considered to be what is referred to as a 'gentlemen's agreement'. With over 50 years of experience in the construction industry, I cannot recall when any such promises of future benefits ever came to fruition. This has been due to a variety of reasons, including the following:

1. The Employer's personnel involved, albeit that they themselves considered that they had full authority to make such offer, either moved on to other jobs or retired, whereupon the carrot disappeared out of the door with them. Those who took over nearly always claimed not to know about the carrot and, without fail, countered that such agreement was, in any case, not embodied in the Contract; and there will always be wording in the Contract that backs them up.

2. The carrot never saw the light of day, either because economic circumstances changed dramatically or because a rethink of the Employer's future development plans meant that the carrot would no longer be on offer.

3. The Employer's purchasing/procurement rules did not allow the carrot to be delivered in the previously anticipated manner, or new rules were introduced that required a formal and more competitive route to be taken for the carrot to be delivered. I then witnessed all such carrots being delivered to others, sometimes with the Contractor not even being included in the bidding frame for that carrot (despite the Contractor having worked hard for it).

The net result of all those changed circumstances was that the Contractor had taken on additional commitments for which there was no reimbursement coming from the Employer's side. In some such cases, I then observed dramatic financial consequences that completely eroded the Contractor's anticipated profit on the Project. There is, unfortunately, no other way to avoid this situation than to consider that any such 'gentlemen's agreement' offered to the Contractor will eventually prove to be completely worthless. Taking that stand will then require the Contractor to negotiate fair recompense for the extras required by the Employer, with the benefits thereof being properly locked into the contractual provisions.

It is worth noting that, in all cases of carrot-offering I witnessed, the Employer never lost anything. On the contrary, in some cases the Employer made immediate substantial gains as well as also having the potential for future gains. Contractors, of course, never get any of the benefits of future gains from the extra efforts and additional money they put into Projects, except under the Build-Operate-Transfer model. Promises of carrots on the standard EPC Project will therefore generally not make a Contractor fat but thin, the visible impact of that thinning always being on the Contractor's bottom line. A number of the Projects I was involved with suffered in this way. On one particular Project, the Contractor had to say goodbye to an amount that equated to almost 25% of the Project's original value.

4.23   Square Pegs and Round Holes

It is a fact of life that not all management personnel operate in the same way or possess equal ability. Any management team will have people with a wide range of abilities and a variety of different character traits (with a few displaying attitudes that are not so nice as some others). The abilities and qualities of the Contractor's management personnel (Directors, Managers and Sub-Managers) will play a very critical role in determining whether or not the Contractor's Projects will be successful. It will be up to the Project Manager to mould the management group into a cohesive unit, one in which all the members cooperate smoothly with each other (and with subordinate staff too).

A lot more is expected nowadays of the Contractor's Project Management Team than in decades past. Today it is recognised, much more than before, that companies should strive to make the workplace a psychologically safe place. This objective has to start with (and be evidenced by confirmatory action from) the Corporate Management Team. As Osterweil observed, everything that somebody in a management position conveys to his/her staff members, whether by verbal or non-verbal means, is transmitted into the feeling systems of those others and can therefore have a direct impact on their performance.[4] The relevance of 'attitude' within a team is also strongly supported by the findings of Google's 'Project Aristotle', whose researchers (after examining 180 teams and scrutinising 250 variables) concluded that 'who is on the team matters far less than how the team members interact, structure their work and view their contributions … when we are considering these dynamics, psychological safety stands head and shoulders above all the others'.[5,6]

One thing I have observed over the many years that I have been working in the construction industry is that there will always be people who, once they have been placed in a management position, believe that they are there simply to make sure that everybody working under their control does their job properly. Some even believe that actually doing anything other than 'managing' and 'directing' is beneath them. They do not consider that they themselves should have to roll up their sleeves and do any of the work that their subordinates are engaged in. I have a very different take on what a Manager's responsibilities and inputs are, my key point being that all Managers must contribute in good measure towards getting things done efficiently and to a high quality standard. Sadly, I have worked in companies where a certain degree of nepotism and cronyism influenced the decision to place inadequately qualified personnel into positions of high responsibility and, ultimately, I witnessed their management failures. Very many years ago, a good friend of mine (one James Porteous Brown Simpson) made the statement that 'If you try to force a square peg into a round hole, there is a good chance that both the peg and the hole will be damaged'. Subsequently, I observed that what he had stated proved to be all too true.

The best Managers I came across got actively involved in the 'doing' work by helping out on the tough parts, and very often guided/trained others as to how to tackle difficult tasks effectively. When conducted properly and earnestly, the mentoring efforts can be hard work, but the results are nearly always the same; the work is completed better and faster, and the staff are truly appreciative of the learning experience. I also believe that a Manager who is unable to list what his/her top priorities for the week are (activities that will aid the work flow) is a waste of space and holding things back, not helping move things forward. A good Manager will always maintain a weekly prioritised 'To Do' list and use it to help him/her focus on what the serious issues are that need to be attended to. Distractions can easily occur in the construction world, and it is more than likely that any Manager who tells you that a 'To Do' list is not necessary for them is only doing half a job.

---

4 Osterweil C. (September/October 2017). Brain food (how brain functions can affect the management of construction Projects). *RICS Construction Journal*, pp. 16–17.
5 Schneider M. (19 July 2017). *Google spent 2 years studying 180 teams. The most successful ones shared these 5 traits.* https://www.inc.com/michael-schneider/google-thought-they-knew-how-to-create-the-perfect.html (accessed 22 November 2018).
6 Thomson S. (9 December 2015). *Google's surprising discovery about effective teams.* https://www.weforum.org/agenda/2015/12/googles-surprising-discovery-about-effective-teams (accessed 22 November 2018).

In addition to paying attention to creating a harmonious and contented workforce within their respective Departments, Department Managers must make themselves fully aware of the risks inherent in undertaking the work of the Department itself. The Managers must also make sure that they understand the chances of those risks becoming real problems, and then put into place mitigation measures for reducing (or even eliminating) those risks. As Ashkenas states: 'Every Manager is a Risk Manager'.[7]

Further, each Manager must keep in mind that there are many other allied Departments and sub-Departments that have Managers and Sub-Managers in charge of the work activities beyond the Engineering, Procurement and Construction Departments (whose personnel, of course, undertake the principal activities on an EPC Project). This means that all Managers are obligated to think beyond the walls of their own Department/sub-Department and ensure that the interfaces of their respective work with the work of all those others are performed in a timely manner in both directions (both to and from those others).

In view of the above, it is therefore vital that the Contractor ensures that its selection process for its management personnel is robust and designed to weed out the also-rans, since anything less will run the risk that its Departments are not functioning as effectively, productively or efficiently as they could be. The inevitable result of employing inadequate Managers will be that money will be lost in a variety of directions, simply because such Managers will not be aware of where the inefficiencies are within their respective Departments or what is causing them. The Contractor's overriding consideration must therefore be to establish precisely what the shape of the management hole is, then select a Manager of the right shape to peg neatly into that hole; forcing a square peg into a round hole must be avoided wherever possible. Having seen where Projects often go wrong, I am of the opinion that the extra cost of employing good Managers will generally be more than compensated for by the more effective and productive working that worthwhile Managers engender.

4.24 Failure to Check the Margin

As can be seen from the foregoing, if attention is not paid to identifying the full costs of the implementation work during the bidding phase, it is possible for a Project to be financially handicapped before the Contract has even been signed. It is therefore imperative that the general pricing margin is large enough to cover the typical level of minor inadequacies that will inevitably be found later in the basic costs computed for the Project.[8] Any costing oversight can create the scenario where financial losses for the affected element of work become inevitable. However, usually there is a good chance that the pricing margin built into the other work elements will save the day, even though the overall level of profit may be a little reduced from the initial expectations.

---

7 Ashkenas R. (3 May 2011). *Every manager is a risk manager*. https://hbr.org/2011/05/every-manager-is-a-risk-manage.html (accessed 28 October 2018).
8 Pricing margin as used here refers to the difference between the bidder's asking price and the bidder's costs for producing/delivering the completed facility. Margin is usually expressed as a percentage of the selling price.

On the other hand, a major error in the costs for a significant component could wipe out all chances of the Contractor making a profit from the Project, despite the overall pricing margin being adequate. I therefore consider that it is the responsibility of the Board of Directors (BODs) to ensure that (i) the bid pricing for Projects can be compiled in the best possible manner and (ii) the pricing margin employed stands a good chance of the overall anticipated financial results being delivered. This entails the Corporate Management Team using competent personnel (who utilise tried-and-tested computer technology) that will enable worthwhile checking of all the components of the bid costings to be performed easily and meaningfully. The aim of the cost checking exercise must be to ensure that the costings against which the pricing margin is being applied are adequate; there can be no glory in being awarded a Project where it is almost impossible to make a profit.

However, no matter how diligent the BOD and the Corporate Management Team have been, or how well priced a Project may be, poor implementation work very often sees the profit considerably diminished from what had been expected, despite the costings and the pricing margin having been adequate. The overall responsibility for ensuring that the implementation work for a specific Project yields at least the anticipated profit rests with the relevant Project Manager (supported by the Project Director). Having said that, the Project Manager is very dependent on the Managers comprising the Project Implementation Team to control the work in their respective Departments. And that is where things often go wrong, usually as the result of a series of small mismanagement issues that are gradually compounded over time. The primary objective of the remainder of this book is therefore to identify where the largest losses can occur in the implementation work, and how to avoid such losses occurring. Nonetheless, pricing the bid for the right management team to be provided in the first place is absolutely essential, because the better the quality of the Managers, the more likelihood there will be of losses being contained/controlled, thereby ensuring profitability.

Chapter 5

Overseas EPC Project Preparatory Work

5.1 Critical Path Identification

Without the Engineering work being completed on schedule, timely completion of the Procurement activities would be compromised, as too would the later Construction activities. Similarly, if the Procurement work is not progressed in a logical and timely fashion, then the progress of the Construction activities may suffer later, due to materials, goods and equipment possibly not being available in time to undertake work that sits on the Project's 'Critical Path'.

The words Critical Path are easy enough to say, and the phrase is bandied about a lot when discussion takes place about a construction Project's progress (or lack of it). However, I find it amazing how few people in the construction industry can provide an off-the-cuff worthwhile definition of it. Many different versions can be found in print, not the least important of which are the following concise (but nonetheless very useful/informative) definitions provided by the Association for Project Management (APM):

'A sequence of activities through a network diagram from start to finish, the sum of whose durations determines the overall duration. There may be more than one such path.'[1]

Alternatively: 'This is the chain of activities connecting the start of a network with the final activity in that network through those activities with zero float. There may be more than one critical path in a network.'[2]

Additionally: 'The path through a series of activities, taking into account interdependencies, in which the later completion activities will have an impact on the project end date or delay a key milestone.'[3]

1 Association for Project Management (2012). *APM Body of Knowledge*, 6e. Association for Project Management, Glossary, p. 236.
2 Association for Project Management (2015). Planning, Scheduling, Monitoring and Control. Association for Project Management, Glossary, p. 332.
3 Association for Project Management (2013). Earned Value Management Handbook. Association for Project Management, Glossary, p. 168.

I have been involved over the past 20 years or more in teaching young foreign engineers about both project controls and contract administration work. From the outset, I found there was a lot of confusion about exactly what the Critical Path was. Specifically, the plethora of different definitions that exist did not help the situation. Some definitions, unlike those of the APM, were convoluted mind-twisters. However, quite a few of my students to whom programming was new had trouble applying the above APM definitions to even the simplest of programmes presented to them for teaching purposes. Importantly, many failed initially to grasp the fact that the Critical Path referred to the longest overall duration of interconnected activities. After discussion over the years about the problem with, amongst others, a fair number of senior planning/scheduling experts, I eventually arrived at the following definition for the term Critical Path:

> *A sequence of logically connected, interdependent activities that have no float within a networked project implementation programme, from the planned start date through to the anticipated finish date, and which results in the longest overall duration for achieving completion of the project.*

As I intimated above, that definition is an amalgamation of inputs from various discussions that took place over the course of a few of years with a number of people more acquainted with scheduling that I am, as well as from various surveys I conducted amongst different groups of colleagues over the years. My thanks go to all those who collaborated with me to help develop the definition into a useful teaching tool, which I have found prompts very active discussion when it is being dissected, analysed and put to use.

I have lost count of the number of Projects where the Critical Path was constantly being referred to by people outside of the Project Controls Team who did not actually know what the critical activities were. Many had not attempted to study the Project Schedule. That meant that they could not tell you where the Project's progress truly was in relation to the Critical Path and whether or not the Project was in delay. The primary reason for that seemed to be because a Project Schedule can easily involve thousands of activities that very often are not brought together and displayed in an organised fashion that will lend itself to easy analysis. My observations also led me to conclude that most construction professionals would have a hard time producing a meaningful Project Schedule even in simple bar chart format. As for determining for themselves which activities sit on the Critical Path, I have seen that most construction professionals would not know where to start. It is therefore easy to see why the correct concept of the Critical Path eludes many more than you may care to think about.

To be fair to construction professionals, it must be said that effective Project planning is far from an easy task, since it requires a very detailed knowledge of the full spectrum of Project activities, tasks and interfaces/relationships, as well as the resources necessary to undertake them. A tremendous amount of iterative work goes into achieving a worthwhile Project Schedule, and it is usually undertaken by a group of dedicated experts over an extensive period of time. Those experts also have to make a large number of assumptions that, very often, they keep to themselves. Added to that, due to the fact that the Project Schedule needs to be issued quickly to be of any real value, many of the required logical links may be missing because of the rush to issue it.

It is that very same complexity that causes some Managers to eschew the Project Schedule and try to conduct their Department's work without paying it too much attention. However, because of the high level of importance that maintaining the Project Schedule occupies in determining the success or otherwise of a Project, I deal with this topic in much more depth in Section 6.6 (The Project Controls Department) and also Section 6.7 (The Planning/Scheduling Team).

5.2 Setting Up Contractor's Administrative Systems

Following the official signing of the Contract, the Employer will mobilise its Project team responsible for monitoring the performance of the Contractor and the progress of the Project. It is common for the Contractor's Engineering and Procurement activities to be carried out from the Contractor's main offices in its home country or in a safe/secure overseas location (as distinct from the sometimes less hospitable or more hostile countries in which the Project is to be built). It is also generally the Contractor's responsibility to provide the work area in that same office to the Employer's Project team (in line, of course, with the specific Employer's requirements and any special conditions written into the Contract). This arrangement makes for easier coordination and communication between all those involved in the Engineering and Procurement work. However, it can take a lot of time re-arranging workspaces to ensure privacy for the parties, as well as for ensuring that all IT communications are in place for all workstations (including access to the Internet, the computer servers and the Electronic Document Management System being used for the Project). Since most EPC Projects are on a very tight time-line, ideally this preparation work needs to be well under way in advance of the Contract start date.

5.3 Determining Appropriate Management Structure

As I explained in Section 1.2 (The Book's Content and Structure), I made the decision to adopt a specific EPC Project organisational management structure for the purpose of allocating risk responsibilities to specific individuals. Appendix C (EPC Project Management Team Organisation Structure) contains the diagram that identifies that structure. From personal experience, I know that it is not possible to provide an organisational structure that will satisfy everybody (not even amongst the management personnel working on the same Project), so I am not concerned about nailing my colours to the mast for the purposes of this book. Despite the fact that any given Contractor may employ a significantly different organisational structure from the one I have employed, the arrangement I have opted for is the one on which the comments and advice in this book have been predicated. It is a structure very similar to that I have seen in use on quite a few occasions, although I have to admit that sometimes it was changed markedly from Project to Project within the same company. That was usually because the nature of the work or the work scope/content required a somewhat different approach.

The organisational management structure I have adopted should therefore be seen as a guide to how the wide range of participants could be organised, not as a structure that is written in stone. Nonetheless, when it comes to allocating the roles and responsibilities of the individual Managers, I tend to be far less flexible in my thinking. The reason for that is primarily because I have been involved in too many situations where things went awry because certain Managers wrongly thought that a key management function or activity was somebody else's problem, not theirs.

A good deal of the remainder of this book therefore concentrates on the roles/functions of the management personnel involved in a typical EPC Project, as well as on the Departments themselves. This is because I believe that the head person in each Department should be the one who is required to identify and manage the risks inherent in the work being conducted by the various Disciplines he/she is responsible for inside his/her Department.

In order to be able to focus on what the risks for the work of the individual Managers are, I therefore considered that it was first necessary to accurately define the responsibilities for each of those individual Managers. The rationale behind this line of thinking is that, wherever/whenever a responsibility is placed upon somebody, there is a risk that something may occur to stop that responsibility from being properly fulfilled or faithfully carried out. For the foregoing reason, in Chapter 6 (Project Roles, Functions and Responsibilities) I have opted to set down the major responsibilities of each individual Manager first, before launching into my own view as to how each individual Manager should manage the risks in his/her own Department.

In addition, there is one particular party outside of the Contractor's organisation that will have a major impact on the Contractor's activities. This is the entity responsible for acting as the Employer's eyes and ears when it comes to monitoring the Contractor's implementation work. That entity may be a team of people from within the Employer's organisation or it may be an independent party, the latter more commonly being referred to as the Project Management Consultant (PMC). The PMC has certain responsibilities towards both the Employer and the Contractor, and how those responsibilities are fulfilled can often prove critical to the success of the Project from the Contractor's perspective. Chapter 6 therefore commences with explaining the primary functions, duty and loyalty of the PMC before moving on to deal with the entities within the Contractor's own organisation charged with the responsibility for handling the Project implementation work or providing support thereto.

Chapter 6

Project Roles, Functions and Responsibilities

6.1 The Project Management Consultant

6.1.1 Primary Functions

Many Employers do not have the necessary in-house expertise to handle the implementation of their own major construction Projects, especially for large-sized Engineering, Procurement and Construction (EPC) Projects. They therefore frequently opt for appointing a Project Management Consultant (PMC) organisation to handle that work for them. Due to the fact that they are often simpler in nature, Design-Build Projects do not generally have a PMC appointed, and the Project is usually administered by an Architect or an Engineer. The PMC for an EPC Project is engaged and paid for by the Employer, and does not form part of the Contractor's Team in any way. However, it is very important for the Contractor to understand the precise role the PMC is required to play and how the attitude and actions of the PMC's members can have a great bearing on the how the risks faced by the Contractor pan out.

The role of a PMC can extend to providing appropriate teams of personnel for carrying out all of the following functions:

1. Reporting to the Employer on all aspects of the Project's implementation.

2. Conveying the Employer's decisions to all other participants in the EPC Project.

3. Preparing drawings and other documentation to show the basic requirements for the Project (sometimes, such as in the FIDIC EPC-orientated contracts,[1] referred to as the 'Employer's Requirements'). *(It is not at all unusual for a separate PMC group to undertake this task, with another PMC taking over either before the EPC bidding process commences or after it has been completed. Such arrangement helps avoid the possibility of a conflict of interest arising in the event that the Contractor raises claims against the base documents prepared by the first PMC.)*

4. Organising and handling all activities involved in the EPC bidding process, the award of the Project and the signing of the Contract.

---

1 FIDIC – Fédération Internationale des Ingénieurs-Conseils. See Section 1.3.

5. Reviewing the Contractor's Deliverables and monitoring the Contractor's implementation and construction work activities to ensure full compliance with the Employer's requirements.

6. Ensuring that all material and documentation required to be handed over to the Employer's Operator is complete in all respects and suitable for the safe operation and maintenance of the completed facility.

In essence, the PMC is required to function as the Employer's eyes and ears, while at the same time being responsible for drawing up the detailed listing of the Employer's requirements in the form of drawings and specification-type documents. However, in order to safeguard the Employer's interests, most PMCs are severely restricted in the extent to which they can commit the Employer without formal (usually written) approval from the Employer. Such restrictions very often mean that the PMC cannot approve any extension of time for completion of the Project or agree any increase of the Contract Price.

6.1.2 PMC's Duty and Loyalty is to Employer

The Contractor must remember at all times that the PMC's personnel work for the Employer, not for the Contractor. The PMC is therefore under no obligation to act impartially towards the Contractor unless the Contract has specific wording covering such an obligation. However, legislation in the country where the Project is to be built (or in the country where the courts of law will have jurisdiction over the interpretation of the Contract) may specifically call for such impartiality. I suggest that the almost automatic lack of impartiality displayed by many PMCs arises as a direct result of the PMC being reliant on the Employer for future work. This means that the principal aim of the PMC will be to show the Employer how valuable the PMC's services are. Sometimes that involves the PMC being very hard on the Contractor, almost to the point of being unfair. In defence of PMCs, it has to be remembered that Contractors must fight their own battles; it would not be seen in a good light by the Employer if the PMC were to be perceived as standing in the Contractor's corner too often.

The PMC's enthusiasm for doing a good job for the Employer often results in the Contractor suffering unexpected delays, frequently due to the over-zealous checking conducted by the PMC's staff. This is nowhere more clear than when it comes to the PMC's reviewing of the Contractor's Deliverables. That is where the PMC's late granting of approval often causes major delay problems for the Contractor. However, most EPC Projects have very rigorous requirements for the outputs required from the completed facility (the 'Guaranteed Performance Outputs'). The entity that carries the sole responsibility for achieving the facility's outputs is the Contractor, not the Employer (and certainly not the PMC). I therefore consider that PMCs ought to be prepared to exercise a larger degree of discretion, and look for ways/reasons that would allow them to grant necessary approvals quicker. Unfortunately, however, I have observed that PMCs exercising such discretion seems to be more the exception than the rule (and, all too often, as a result of a 'jobsworth' mentality). The Contractor should therefore be prepared for the PMC's personnel to be a lot slower in getting things cleared than the Contractor had been hoping for. That is another reason why there ought to be adequate additional time allowance built into all key activities in the Project Schedule (although, too often, I saw that such allowance proved to be inadequate).

6.1.3   PMC's Different Take on Time

I have noticed that many PMC members do not seem to consider it their duty to help save time for the Contractor. I surmise that this may be due to the fact that the PMC's remuneration is usually earned through supplying personnel on a time basis, which means that saving time is not usually a priority for the PMC's own work. Thus, if the time for completion of a lump-sum EPC Project gets extended through no fault of the PMC, then the PMC will generally receive greater remuneration. The fact that this applies for the Contractor only under reimbursable contracts does not seem to register with the PMC's personnel. It is therefore hardly surprising that many of the bidding documents prepared by PMCs make every 'transgression' of the Contractor in its Deliverables a reason for insisting on a resubmission of the corrected document. Such resubmissions must then go through exactly the same review process as the previous submission, yet itself be subject to rejection again and therefore require a further re-review. If the Contract provisions have not been suitably amended to stop this iniquitous process, then it will take a lot of fighting after Contract signing before the PMC will relent and agree to a more reasonable approach to the reviewing of Deliverables (if at all).

To reduce the likelihood of Deliverables being rejected and made subject to repeated re-reviews, the Contractor must ensure that its teams responsible for the preparation of the Deliverables are competent enough to do a professional job at all stages. Even then it will usually be found that there will be too many instances of unreasonable rejection occurring. That will often be due to well-meaning PMC personnel who think that they are protecting the Employer's best interests by constantly 'pushing the envelope' with the Contractor. Very often, it will be found that such people seem to have great difficulty in accepting that they are not entitled to practise (and insist upon) their 'preferential engineering' solutions. Because we are facing a situation here where human pride and endeavour are involved, this scenario will inevitably arise on every EPC Project at some point. However, if the Contractor can demonstrate the unreasonableness of the PMC's actions by virtue of the good quality of the Contractor's Deliverables, such an attitude can be stopped in its tracks as soon as it rears its sinister head. It requires the Contractor to be very firm because, if such a situation is allowed to continue, it will simply become increasingly more difficult to stop, which may then lead to huge delays that can never be recovered from.

Obviously, it is best to start with a tactful approach when dealing with the foregoing situation. The major problem the Contractor will face is that the majority of PMCs seem to think that it is not their duty to avoid delaying the Contractor, and that the Contractor is under an obligation to catch up all delay caused by the PMC, whether or not the PMC's action or inaction is unreasonable. If it becomes necessary for the Contractor to throw down the gauntlet in an effort to win in this unfortunate situation, then it should be undertaken by writing a very carefully worded letter direct to the Employer (copied to the PMC) to complain about the PMC's unreasonable behaviour.

My advice regarding the foregoing is that writing directly to the Employer should only be done after the PMC has been verbally informed that this will be the course of action that the Contractor will take. Of course, the Contractor must ensure that the facts of the matter support the Contractor's position, and that all therefore comes back to the requirement that the competence level of the Contractor's documentation teams must be adequate. After all, it would be unreasonable to expect that the PMC should approve the Contractor's documentation if it is inadequately prepared and which left too many questions unanswered (and which, sadly, I have also seen occur too many times). A balance therefore has to be struck regarding what constitutes an acceptable standard for the Deliverables, but the ball is very much in the Contractor's court to begin with. It should not be left to the PMC to have to fight to get a reasonable standard established for a Deliverable.

6.1.4 Employer's Direct Personnel in Lieu of PMC

Similar problems can arise if a PMC does not exist but, instead, the Employer's Team for monitoring and administration purposes is drawn from among the Employer's own direct internal personnel. In such a case there will inevitably be a far greater feeling of loyalty of those persons towards the Employer than if a PMC team is involved. However, difficult topics can usually be more easily elevated to the Employer's top management team when there is no intermediary in place (as there is when a PMC is engaged). For that reason, it is usually easier for decisions to be made in a more timely manner where the Employer's own people are dealing directly with the Contractor's personnel.

I have often heard it said that the primary objectives of the Employer and the Contractor make the relationship between the two inherently adversarial, and many seem to believe, with good reason, that it is like that for every Project to one degree or another. That theory would certainly be proved true if an Employer were to set out with the intention of acquiring a first-class facility at the cheapest possible price, stopping at nothing to achieve that. It would be equally true if a Contractor were to set out with the aim of continually cutting corners in quality in order to make a bigger profit. However, I have not seen such extreme attitudes when dealing with well-established, reputable companies, neither from the Employer's side nor from the Contractor's side. I therefore consider it wrong to believe that future transactions between the two contracting parties will be conducted on an adversarial basis most of the time, provided that business is conducted in an atmosphere of mutual respect.

My line of thinking is premised on the fact that the Employer will always be fully aware that the completed facility represents a valuable asset that will generate considerable income once it is put into operation, and that helping the Contractor to achieve timely completion of a Project is therefore essential. If the Employer does anything that will negatively impact the Contractor's progress and/or profitability, it will most certainly not help achieve timely completion of the Project. Likewise, a reputable Contractor fully understands that its future reputation depends on it doing a first-class job today, and that cutting corners just to make more money would be one of the quickest ways to destroy its good reputation.

In view of this, both parties should make every effort to form bonds of friendship and trust between the two entities prior to Contract signing, and commit to open and honest dealings at all times, based on the principle of 'One Team, One Objective'. That objective should be to get the Project completed by the agreed-upon time, to the agreed-upon quality and for the agreed-upon price. If the Contractor fails to develop this culture of partnering, it runs the real risk of the inherently adversarial scenario kicking in, which would benefit neither party in the long run. This principle of the Contractor forming a good relationship with the Employer's own internal team members applies even where a PMC has been appointed for Project administration purposes. That would then help tremendously whenever the Contractor needs the Employer's help to resolve problems unnecessarily caused by the PMC.

6.2 The Board of Directors

6.2.1 Primary Functions

All large contracting organisations will, at the very top level, be run by a Board of Directors (BOD). Although the BOD does not have any direct responsibility for the day-to-day running of the EPC Projects the company is awarded, the BOD is most certainly responsible for the following in regard to all the company's Projects:

(i) ensuring that adequate support is provided for each and every one of the Contractor's Projects, such as will maximise the chances of their success;

(ii) monitoring everything that is happening on each individual Project; and

(iii) stepping in and insisting that appropriate action is taken by the Corporate Management Team whenever it is apparent that a Project (or its key personnel) may be heading in the wrong direction.

I am referring above to the overall responsibility for managing the Contractor's company. The responsibility for running the daily operations of the Contractor's business is undertaken by the Corporate Management Team, headed up by the Chief Executive Officer (sometimes called the General Manager, amongst other titles). The BOD must take the responsibility for ensuring that suitably qualified personnel are appointed to the Corporate Management Team. The BOD will rely on the Corporate Management Team to do everything necessary to run the company's operational activities satisfactorily and to provide ongoing support to the Projects.

The Chief Executive Officer will be responsible for keeping the BOD fully informed if anything is not going according to plan regarding all aspects of running the daily operations, including all Projects. Since I regard the Corporate Management Team as an extension of the BOD (there to work according to the BOD's wishes), I do not make any further mention in this book about the roles and responsibilities of the Corporate Management Team or its individual personnel unless I feel it is necessary to do so.

6.2.2 Ethical Compliance Responsibilities

Managing a construction business is very different today than it was just 10 years ago, especially for organisations operating in an international environment. This is because much tighter regulations now require organisations to employ stricter policies for the running of businesses and keeping records, especially of financial transactions. In today's business environment, one of the primary duties of the Contractor's BOD (alongside successful business development and ensuring that Project implementation activities can be conducted smoothly) is to ensure that the company itself is managed in a professional manner. This requires running the business in accordance with the latest applicable administrative and financial accounting rules and regulations, especially in respect of:

   (i) vetting staff before hiring them,

   (ii) conducting 'third party due diligence' when dealing with other organisations (including Subcontractors and Vendors), and

   (iii) adhering to applicable anti-bribery, anti-corruption and anti-money laundering legislation (such as the USA's 'Foreign Corrupt Practices Act'[2] and the UK's 'Bribery Act, 2010'[3]).

An earlier piece of legislation that changed the requirements for how financial accounting is to be managed and reported on was the 'Sarbanes-Oxley Act' of 2002,[4] which was enacted as a reaction to a number of major corporate and accounting scandals, including Enron Corporation and WorldCom. That Act specifies the responsibilities of a public corporation's BOD, and it added criminal penalties for certain misconduct. It also required the US Securities and Exchange Commission to create regulations to define how public corporations are to comply with the law. Further, it revised previous sentencing guidelines, and it strengthened penalties for manipulation, destruction or alteration of financial records, while providing certain protections for whistle-blowers.

The BOD must therefore ensure that policies and procedures are put in place that then have to be followed by all the organisation's employees (which includes all the Contractor's Directors as well as all its Managers and staff members). The BOD must also make sure that effective monitoring and control mechanisms are established to check that the organisation's policies are being followed and applied properly. That is usually achieved via a Compliance Department, whose members will be tasked with trying to identify weak internal controls that need to be strengthened, or else risk a damaging non-compliance occurring at some point in the future. The BOD, through its Compliance Department, is therefore required to establish a sound 'compliance' programme to create an environment in the workplace where high standards of ethics are the norm, and where the necessary monitoring and control mechanisms will result in ensuring

---

2 The Foreign Corrupt Practices Act of 1977 (FCPA): a United States of America federal law that prohibits bribery of foreign officials and also addresses accounting transparency to avoid being able to hide such transgressions. Any company anywhere in the world dealing with a USA company that breaches FCPA rules can be penalised by the USA courts.

3 The UK Bribery Act 2010 covers much the same topics as the FCPA and in a very similar way. However, while the FCPA makes a distinction between small and large bribery payments, facilitation payments of any type or magnitude are prohibited under the UK Bribery Act.

4 This is the more commonly used shortened form (nickname) of an Act of Congress enacted as federal law in the USA, which was introduced with the specific purpose of protecting investors by improving the accuracy and reliability of corporate accounting.

that sound business judgement is applied at all times. Taking such steps should thereby provide a valid defence in the event of any unforeseen non-compliance issues arising either internally or with a third party.

Ensuring that compliance policies, rules and regulations are being followed requires personnel to be available who are specifically dedicated to the task of monitoring compliance, and they must have appropriate checking mechanisms in place to be able to spot anomalous financial transactions. The BOD has the responsibility for making the funds available for those monitoring and checking activities to be performed properly. Failure to do that runs the risk of the Contractor being excluded from bidding for Projects for such entities as the International Oil Companies (IOC's), since their due diligence checks would almost certainly reveal that the Contractor does not operate a sound compliance management policy. In such a situation an IOC may decide to drop the Contractor from its approved list of potential bidders. This could occur because the IOC would otherwise also be at risk of being found guilty of non-compliance if, subsequently, the Contractor or one of the Contractor's Subcontractors is declared guilty of a significant compliance violation.

It is also insufficient for the Contractor simply to have a compliance management policy in place, even if they are supported by a worthwhile set of written internal rules and regulations. Such things must be complemented by and followed up with rigorous training of all personnel affected by such rules and regulations, where the compliance training is required to be formal and recorded. Failure in regard to this matter would make the Contractor liable if any of its employees violated the external compliance rules and regulations to the extent that legal proceedings became necessary. That in turn would tarnish the Contractor's good reputation and destroy goodwill, thereby making it more difficult for the Contractor to get onto the bid lists for future Projects. The BOD must therefore take a keen interest in the activities of the Compliance Department and insist on getting regular reports of the compliance monitoring and compliance training progress. The BOD also needs to demonstrate that (i) it has responded effectively where weaknesses in the internal control mechanisms were revealed in the Compliance Department's reports and (ii) it has seen to it that appropriate changes to overcome the problems identified have been implemented.

6.2.3   Shop Floor Compliance Problems

There are of course many other corporate risks that the BOD will need to consider and provide mitigation plans for but, of all the corporate risks, the one where a Project's personnel have the ability to cause a major upset is with *ethical non-compliance occurring at the shop-floor level*. This is something that the BOD must be constantly aware of and also be seen to take the lead for within the organisation. That is best done by the BOD committing to and conducting the business in a fully compliant manner in all respects with regard to administrative, legal and personal ethical requirements. In particular, the BOD must ensure that the company has adequate policies and training in place to cover such issues as anti-bribery, anti-corruption and anti-money laundering. Further, the BOD must ensure that adequate employee training schemes are in place regarding such compliance issues and that the BOD is plainly seen to be seriously concerned about uncovering non-compliances.

As it happens, the construction industry is rife with opportunity for those bent on employing corruption to further their own ends. In fact, a figure of USD 1 trillion per annum is quoted by the World Bank as the amount paid in bribes across all businesses globally,[5] where the construction industry is one of the largest business endeavours on the planet. The risk the BOD therefore runs by not taking the lead in the fight to combat non-compliant behaviour is that employees may try to take advantage of the situation due to the lack of internal controls. However, violation of international compliance rules and regulations by an employee that leads to criminal charges may also be considered to be an equal transgression of their employer. The employer's defence is to be able to show solid proof that the organisation took substantial steps to try to make sure that its employees were fully aware of what constitutes unacceptable ethical behaviour. If an organisation is convicted of not having taken appropriate steps for stopping non-compliances, then the fall-out in the business community could be substantial and, in the worst-case scenario, could destroy the Contractor's business completely.

The best way for a Contractor to avoid the risk of employees engaging in corruption is to reduce the amount of ad hoc paperwork that needs to be done. That can be achieved by implementing enterprise-wide computerised management systems for as many tasks as possible. At the very least, I consider that it should be done for the purchasing and payment functions (such as via SAP – 'Systeme, Anwendungen und Produkte in der Datenverarbeitung' – also known as 'Systems, Applications and Products' in its simpler English version).[6] Utilising such systems will enable clean, open and transparent business to be conducted, and the data can be set up to be visible only to authorised personnel strictly on a 'need to know' basis, thereby enabling commercial confidentiality to be maintained. The system employed can be set up so that automatic flagging up of possible problem payments can be highlighted by sending email alerts to key personnel responsible for approving such payments. By contrast, manual payment systems are wide open to being abused. This especially occurs in manual payment environments where a robust audit trail is not evidenced, and paperwork can thus be deliberately lost, misplaced or adulterated without being detected easily.

6.2.4 Corporate Directives for Limiting Compliance Transgressions

Another critical step that the BOD should insist upon is to restrict the involvement of each individual Department's personnel in undertaking the direct work of the other Departments. For example, only those in the Procurement Department should be authorised to make contact with the bidders once the Requests for Quotations and Bid Enquiry Documents have been issued to prospective Vendors and Subcontractors. This means that, if revised documentation of any sort has to be issued to the bidders, it should only be done via a Bid Bulletin issued by the Procurement Department.

Further, the Engineering Department's personnel must not be allowed to decide unilaterally to change the Procurement Strategy. For example, if formal management

---

5 United Nations (10 September 2018). *Global cost of corruption at least 5 per cent of world Gross Domestic Product, Secretary-General tells Security Council, citing World Economic Forum data.* https://www.un.org/press/en/2018/sc13493.doc.htm (accessed 18 September 2018).
6 The name originally included the words 'in Data Processing', but these were subsequently dropped somewhere around the late 1980s.

approval has been given to the strategy that the Building Works Subcontractor will be responsible for the Electrical Installations work, the Engineering Manager cannot alone later make the decision that all luminaires and distribution boards should be pulled out from the Building Works Subcontract and given to a particular Vendor. This is sometimes done in good faith, on the basis, for example, that it will meet the Employer's requirement that each item of equipment and all accessories must be obtained from a single source (to reduce the need to hold multiple spare parts in stock). Such changes in strategy need to be discussed at a higher management level, with solid reasoning formally presented, because they are likely to have adverse knock-on effects that need to be thought through properly.

The primary reason for introducing such levels of control is to make each Department Manager responsible for trying to stop corruption activities within his/her specific Department. It needs to be borne in mind constantly that the more that one Department's personnel are allowed to cross over into another Department's territory, the more the chances both for mix-ups (confusion) to occur and for corruption to flourish. The problem here is that the BOD may consider that the Departments will be automatically self-policing regarding such policies, but I have observed that it is rarely the case. It is therefore incumbent upon the BOD to consider where things can go wrong further down the line, to put the requirements of the BOD into writing, and then issue the resultant documents as corporate directives to the Corporate Management Team.

6.2.5 Speeding Up Financial Decision-Making

On mega Projects, the number of different Purchase Orders to be placed and Subcontracts to be let can run into many hundreds. This is where money can all too easily be lost very quickly and in large amounts. However, this is also where a great deal of time can be lost if the Contractor's Head Office wishes to exercise control by imposing a long chain of command and convoluted authorisation requirements before Purchase Orders or Subcontracts can be issued. There is a delicate balance to be struck here, between making sure that critical orders are not delayed in any way while also ensuring that the expenditure is being tightly controlled. Quite naturally, the BOD will want to ensure that adequate controls are in place regarding expenditure, since taking the eye off that ball will inevitably see it being dropped.

The foregoing situation is brought about because, nowadays, the funding for most Projects comes not from advance payments from Employers (as it once used to) but from the Contractor's Head Office. The Chief Finance Officer will therefore need to know beyond any shadow of doubt that the projected cash flow requirements are not being exceeded or, alternatively, be given early warning as to the extent that the financial projections are going off the rails. This can lead to very restrictive ordering and payment procedures being put in place, in which Head Office authorisation must be given to each and every order placed. However, that approach will inevitably result in the Project's progress being slowed down severely, leading to unacceptable and, from the Employer's standpoint, inexcusable delays.

To avoid that clumsy situation, my suggestion is that it would therefore be best to establish a separate Project Executive Committee (PEC) for the larger Projects, composed of at least five people. Such personnel could be the Project Director (PD), the Deputy Project Director, the Project Controls Manager (PCM), the Project Finance Officer and the Deputy Project Finance Officer. The PEC should be charged with the responsibility for vetting the recommendations for the placement of Purchase Orders and Subcontracts, and for checking (i) that all the Procurement rules have been followed faithfully (particularly the competitiveness requirements), and (ii) that value for money is being obtained as much as possible. Such an arrangement would also go a long way towards helping the Contractor ensure that the now very necessary international 'due diligence', 'anti-bribery', 'anti-corruption' and 'anti-money laundering' requirements are being implemented both properly and thoroughly. Of course, the rules for the operation of the PEC should also be clearly set down, including the monetary limits of the PEC's authority for placing orders without first having to seek approval of the BOD.

Because of the obvious conflicts of interest involved, the people who most definitely should not be proposed to form the PEC are the Project Manager, the Engineering Manager, the Procurement Manager, the Contract Administration Manager and the Subcontracts Manager. However, any or all of those persons could be invited to attend the weekly (or emergency) meetings called by the PEC for the purpose of deciding which orders to place based on the recommendations of the appropriate members of the Contractor's Project Implementation Team. Such additional personnel should attend in an advisory category only; they should not be granted any 'voting rights' or be part of the final decision-making process. They should be there simply to offer background advice/information and answer pertinent questions regarding the appropriateness of the recommendations received by the PEC.

6.2.6 Ensuring Veracity of Reporting

One area where I have observed big problems occurring, especially on mega Projects, is with regard to the accuracy of financial reporting by the Project Implementation Teams, which I have almost always found to be over-optimistic. This has generally stemmed from a combination of the misreporting for key issues, such as:

(i) overstating the actual progress of the physical work,

(ii) underestimating the amount of the resources and time necessary to complete the remaining work,

(iii) disregarding valid claims from Vendors and Subcontractors, and

(iv) underplaying problems such as late Vendor deliveries and insufficiency of the resources of Subcontractors.

The net result of such misreporting has too often been that, very late in the day, the Project's finances have been determined to be in a much more perilous state than the BOD had been aware of. Perhaps this is what happened in the unfortunate case of Carillion plc's collapse.[7] The question that has to be asked is how such a situation can occur so regularly, and I believe that I have the answer to that question. It is the fear of being

---

7 Pooley, C.R., Pickard, J., and Cumbo, J. (of *Financial Times*) (15 January 2018). *Carillion collapses into liquidation.* https://www.ft.com/content/5ea57733-0c7c-3ccd-9108-6380250c71fc (accessed 6 March 2018).

the messenger who is killed after having delivered bad news (such as being sidelined for promotion, or worse), or fear of being despised as the whistle-blower who 'ratted out' his or her colleagues. Such reticence can be seen at every level. For example, the Lead Engineer will not tell the Engineering Manager that certain critical design documents will be late. When the Engineering Manager eventually finds out that there is a problem, the Project Manager may also not be informed until much later in day (all too often, when it is too late to take action to mitigate the negative effects of that problem).

I have suggested in vain to various management teams that what is required to safeguard the financial interests of a Contractor on a mega Project is to engage an external team of construction-orientated financial experts to continually monitor/audit the Project at a micro level. However, what tends to happen is that too much reliance is placed on the annual fiscal audit, conducted by third parties, to try to pick up on problems when reviewing the company's books at the macro level.

The arrangement I have proposed would overcome the problem of members of the Project Implementation Team having to take direct responsibility for reporting major financial and other problems to upper management. In addition, it would give the BOD far more comfort and confidence in the financial figures. I have suggested that a team such as this should focus on establishing the means and methods of properly measuring the progress of the EPC activities. It should also set up a comprehensive system for monitoring and recording all expenditure, and also be responsible for checking that the budget set for each individual department and work package is being adhered to strictly. The findings of the external audit team should be reported to the BOD on a regular monthly basis, where all major anomalies observed should be fully discussed, and suitable remedies then instituted to correct any problems found.

The objections I have heard about implementing a third-party vetting arrangement of the foregoing nature have been varied. They range from such things as the huge extra cost involved ('It's not in the budget', being the more usual cry) and the possibly debilitating impact on the feelings of the members of not only the Project Management Team but also the members of the Corporate Management Team. I have even heard misgivings about the integrity of the third party's personnel, notwithstanding the fact that all the well-known firms of accountants that handle such matters utilise highly qualified personnel holding certificates issued by Chartered Institutions, including Accountants, Design Engineers, Construction Engineers, Planning Engineers and Quantity Surveyors.

My opinion is that the consequences of not employing a worthwhile external financial audit watchdog for mega Projects will generally lose the Contractor far more than the fees payable for such professional services. However, when such consultants are finally brought in at the micro level, it is generally to sort out the books after a Project termination has been declared or to deal with the fall-out from a declaration of the Contractor's bankruptcy. On the other hand, most of the Projects I have been involved with would have benefited from an earlier third-party external audit team being involved to some extent or other. I also consider that such a mechanism would help keep the Contractor's Project Implementation Team on its toes and thus reduce the possibility of potential or looming hazards metamorphosing into hideously unmanageable problems. I am certain that it would also help tremendously in deterring corruption.

6.3   The Project Director

6.3.1   Standing of Project Director

In an effort to ensure that each major EPC Project will achieve successful completion and meet the Contractor's required business objectives, it is common practice for most Contractors to appoint a Project Director (PD) to sit at the head of and lead the Project Management Team. The BOD will consider that the primary responsibility for ensuring the success of the Project falls squarely on the shoulders of the PD. This is despite the fact that it will ultimately become the responsibility of the BOD if the PD fails to ensure that the requirements of the Employer are fully satisfied.

The BOD must be understood by the PD to be his/her topmost 'customer', whilst not forgetting that the reporting must also be shared with the Corporate Management Team. This is because the authority of the PD to operate as the paramount decision-maker on the Project (as well as to be the face and voice of the Contractor in all matters relating to the Project) ultimately derives from the BOD. There may therefore be restrictions placed on the PD regarding expenditure commitments (by way of prior approval being required from the BOD). Notwithstanding that, in view of the heavy responsibility placed on the PD for successful Project completion, the BOD should empower him/her with full authority for determining, in conjunction with the Project Manager and the Corporate Management Team:

(i) the physical resources necessary for the Project (i.e. the materials, equipment, plant, manpower, construction facilities, etc.), and

(ii) the eventual allocation/distribution of those resources.

6.3.2   PD's Functions and Responsibilities

The functions of and the responsibilities placed on the PD are typically as follows.

- To represent the highest level of authority on the Project, both internally and externally.

- To be overall in charge of and support the entire Contractor's Project Management Team.

- To be the Contractor's primary representative for liaising with and attending management meetings with the Employer.

- To attend the Employer's Project Steering Committee meetings and contribute towards resolving all issues arising therefrom regarding the Project's execution.

- To report faithfully to the BOD as well as the Corporate Management Team, so that the true status of the Project's progress and performance is fully known to those others at all times.

The corporate objectives to be achieved in respect of the Project will be as determined by the BOD, and the reports from the PD to the BOD must cover all aspects of the Contractor's business and interests impacted by the Project. The reports should cover, but by no means be limited to, the following: progress status, financial situation, health and safety performance, quality issues, and local community and environmental matters. To be able to do the foregoing effectively, the PD will be required to keep close tabs on every aspect of the Project's activities. The PD will therefore need to put in place an adequate monitoring and reporting system to ensure that all relevant facts are properly garnered and compiled.

As mentioned above, the support for the PD is ultimately provided by the BOD and, since the PD is the link between the Project Implementation Team and the BOD, the PD is likewise responsible for providing all necessary support to the Project Manager on behalf of the BOD and the Corporate Management Team. The PD will therefore be required to hold regular meetings with the Project Manager and the relevant Department Managers in order to understand in full the true status of the Project at all times. Where it becomes evident to the PD that the Contractor's corporate objectives are not being met (or are in danger of not being met at some point in the future), it will be the PD's responsibility to resolve such issues. Alternatively, in the event that the problems go beyond Project level, the PD must report such matters to the BOD and the Corporate Management Team as quickly as possible.

6.3.3 Good Communication and Quality Reporting

Since the PD is the person responsible for representing the Contractor in front of the Employer, the BOD must give the PD full authority to commit the company. However, although the PD is the person responsible for signing off against any required management approvals, final approval for any major scope changes usually rests entirely with the BOD. This is because major scope changes will increase the liability of the Contractor. Many companies therefore have regulations that require approval of the BOD to be obtained if the value of the Project (the income, not the expenditure) is likely to increase by, say, 2.5% or more or, if the Project is of mega size, by more than a predetermined amount (for example, USD 25 million). This is because significant increases in the value of the Project may requires the Performance Bond value to be increased correspondingly. The PD therefore needs to make sure that an adequate system of checks and balances is in place to ensure that major scope changes requiring higher-level approval do not bypass the BOD's vetting procedure by inadvertently slipping through 'under the radar'.

The wide-ranging scope of the PD's responsibilities means that he/she has no time to get bogged down in the day-to-day workings of any particular Project, and such matters must instead be left for the Project Manager to handle. To ensure that the PD will be able to advise the Corporate Management Team and the BOD properly at any given point in time as to the Project's true status, the PD therefore needs the Project Manager to report regularly in the required structured format. It is not the Project Director's job to compile those reports but to vet them and call for others to amend them if that vetting process uncovers problem areas in the reporting.

The biggest risk faced by the PD is that the information reported to him/her by the Project Manager does not represent the whole truth. Sometimes this may be due to the Project Manager knowingly and deliberately misrepresenting the situation, but more often than not it is owing to inadequate reporting by the various Department Managers. This latter situation can easily occur when a Department's resources are overstretched, which leads to time being taken away from accurately recording what is actually happening. This can occur where staff are diverted from reporting activities to help cope with a Department's heavier than expected workload. In an effort to reduce this risk as much as possible, the PD therefore needs to take the following actions in addition to holding any regular meetings that may be necessary.

1. Hold a joint daily meeting with the Project Manager, the Project Controls Manager and the Planning Manager if/where this is possible/practicable; this need not take long unless something critical occurs. Nowadays, if it is not possible to meet face-to-face due to the Site being located far away from the Contractor's main office where the PD is sitting (which could well be in a different country), this can easily be conducted remotely by the PD, courtesy of such online conferencing facilities as 'Skype for Business'.

2. Chair weekly progress meetings with the Project Manager and all Department Managers present; the aim should be to limit this to no more than an hour, unless critical issues have arisen that will take longer to resolve. I suggest that this is far better conducted in person rather than remotely, but if good videoconferencing facilities are available at both ends, then that could still be a practical and effective way to handle this.

3. Meet separately with each and every Department Manager at about monthly intervals. A face-to-face meeting is essential to get the most out of this type of meeting, preferably at the Site. I have witnessed it proving even more useful where the PD has gone through some elementary training in how to read body language.

The objectives of holding the abovementioned meetings should therefore be twofold:

1. For the PD to keep himself/herself as fully informed as possible about what is actually going on with the Project, thereby providing him/her a means by which to determine the validity or otherwise of the Project Manager's various reports.

2. To keep all the Project's Managers in the picture, helping them to recognise and appreciate that they are an essential part of the Project Implementation Team, and encouraging them to be forthcoming about any problems they perceive before it is too late for the Corporate Management Team to take effective remedial action.

Throughout the entire business world, the reality is that the quality and integrity of reporting will depend very much on the calibre and character of the individuals who make up the management team. No two people are endowed with the same level of commitment to working openly, honestly and with integrity. *It is therefore vital that the PD should get heavily involved in vetting and approving all candidates for every management position on the Project.* The PD should also monitor the performance and capabilities of all the Managers on a continuous basis, and not be afraid to take appropriate corrective action whenever it is found that a particular Manager is failing to conduct him/herself effectively or honestly.

In regard to communicating effectively, the PD should not take any reporting at face value. Instead, he/she should exercise healthy scepticism and question anything that looks the slightest bit dubious. Additionally, the PD should not hold back in testing out the facts surrounding highly sensitive time or costs issues, even if the associated reports seem reasonably sound on the surface. A good tactic is to ask questions and listen to the answers carefully, homing in on any points that do not sound entirely convincing. Only in this way can an acceptable level of confidence be reached regarding the validity of the reporting received by the PD.

The biggest problem I have seen with PDs is their use of intimidation to get the answer they want to hear, instead of encouraging the Managers to give a direct and honest response to a critical issue. Nowhere is this more obvious than when it comes to reporting on how long it is going to take to accomplish critical tasks. However, I have seen a number of Senior Planning Engineers removed from Projects simply because they were too professional to be able to hide their opinions as to the improbability of completing the work on time using just the limited workforce that the PD insisted would be adequate. What made matters worse from my perspective was when the previous incumbent was later blamed by the PD for the planning fiasco that the new Senior Planning Engineer inherited.

My solution to the foregoing problem is simple: *the BOD must make the PDs responsible for having accepted any shortcomings later discovered in their respective Project Schedules.* This would then force the PDs to report any schedule problems to the BOD early on, so that they can be resolved one way or the other in good time. All PDs should remember that killing the messenger who bears bad news is not exactly the best way to proceed. Further, it will most certainly not make the rest of the team feel happy or safe in their jobs. Sadly, however, I have had to work with a few PDs who preferred to operate that way.

6.3.4 Importance of a Good PD

It must be remembered that the PD is the link between the two groups managing the two different sets of risk that the Contractor has to deal with, namely:

(i) the corporate risks (managed by the Corporate Management Team in conjunction with the BOD), and

(ii) those completely different risks that surround the Projects he/she is overall responsible for (managed by the Project Implementation Team).

Both the Corporate Management Team and the BOD are therefore relying on their PDs to feed back all important information on their respective Projects as soon as possible, in order to help avoid a catastrophe occurring, especially one that could lead to the Contractor being declared bankrupt (as happened to the UK's Carillion plc).[8] If a PD becomes too complacent and does not make the effort to dig deeply to ensure that the Project's status is being reported correctly, then that PD may ultimately be guilty of leading the Corporate Management Team and the BOD into a false sense of security.

---

8 Ibid., Chapter 1.

The biggest risk for the Contractor in regard to any given Project is that the right person may not have been selected for the position of PD. What is needed in that role is a true diplomat who is capable of holding his/her own ego in check. The PD must also fully understand and abide by the obligation to be the 'eyes and ears' of both the Corporate Management Team and the BOD. This requires the PD's ability to appreciate when he/she must abstain from communicating in delicate situations until the prior approval of either the Contractor's Chief Executive Officer or the BOD (or, sometimes, even both) has been obtained.

It is not unknown for Projects to go wrong because of personality clashes at the top managerial level. When that does happen, it may be between the two Project Managers representing the Contractor and the Employer. This can occur regardless of the fact that both people are highly competent and generally known as being nice people. Such is the nature of personality clashes and, when one does occur, the PD must be the person to step in and make the right decision as to how to handle it. The choice may be between replacing the Contractor's representative (no matter if he/she is competent) or trying to 'mend the fences' (which is often far harder to do than first meets the eye). Great care must be taken by the Contractor to avoid appointing a PD who may be more inclined to exacerbate the situation or even become part of the problem if a clash of personalities occurs between those others. The PD needs to be viewed as an ambassador for the Contractor, and that requires a person who has a good understanding of diplomacy and patently practises it in their everyday lives.

In closing, it cannot be stressed enough that the PD's role is a vital one. The individual chosen to function as required in that position must be of sound character and thoroughly tried-and-tested by the Contractor beforehand. This is because the PD's performance (or lack of it) can be a big decider in whether or not a Project is going to be as successful as it can be, even where the Project Manager is a first-class, outstanding individual.

6.4   The Project Manager

6.4.1   Primary Functions

The Project Manager (PM) is charged with the responsibility for managing the day-to-day Project work, right through to handover of the completed facility. In some company's the PM is also referred to as the 'Delivery Manager'. He/she is therefore required to supervise and direct all the Department Managers and control the resources allocated to the Project (the manpower/labour, construction equipment and materials), all with the aim of ensuring that the Project is undertaken and completed successfully.

There are three principle areas in which Projects can easily veer off course unacceptably through problems caused directly by the Contractor. If control is lost in any one of those areas, it could well lead to failure of the Project to be completed successfully from the Contractor's perspective. It is therefore vitally important for the PM to pay full attention to managing the following aspects of the Project properly:

1. **The Project Scope:** where the contracted work to be undertaken may increase in scope from what had originally been anticipated by the Contractor. This can occur either without additional compensation being due or without additional compensation having been secured beforehand (even though additional compensation should be due and should have been claimed). An increase in Project Scope can also increase the time it will take to complete the Project. If that time increase is simply down to there being greater quantities of work than the bid pricing had allowed for, then no extension to the time for completing the Project will be granted to the Contractor.

2. **The Project Cost:** where the amount of money budgeted by the Contractor to complete the original Project Scope may be found to be inadequate, but where no additional reimbursement is due from the Employer. This could be down to such things as: (i) the bid pricing having been inadequate (caused by bad pricing or under-estimation of the scope of work involved), and/or (ii) inadequate control of the procurement budgets, poor levels of productivity occurring in too many areas, high levels of materials wastage occurring, materials pilferage/theft, etc. At its worst, this could lead to the Contractor's bankruptcy.

3. **The Project Schedule:** where the contracted length of time in which the original Project Scope has to be completed may become extended due to reasons that are entirely down to the Contractor. This situation could easily lead to the Contractor being required to pay Liquidated Damages for delayed completion of the Project. If that happens, it would generally mean not just that the Contractor's anticipated profit disappears, it could also mean that the Project incurs huge losses for the Contractor.

At the commencement of a Project, all three of the above-mentioned areas can be considered to have a starting 'Baseline' which, if deviated from, may cause the Contractor severe problems, leading to unexpected financial losses. The main problem for the PM is that custody of those three Project Baselines (as well as ensuring that no deviations occur therefrom that are not accepted by and paid for by the Employer), is not directly in his/her hands but sits primarily with others. Those others, set against the problem areas listed above, are: (i) the Engineering Manager, (ii) the Budget Controller, and (iii) the Project Controls Manager. Each of them must keep the PM fully informed, in a timely manner, about any deviations that occur or are anticipated for the area of work under their management. It is therefore vital that the PM ensures that those three Managers in particular fully understand their respective reporting responsibilities, and that they collectively put adequate measures in place to fulfil their respective functions properly. Additionally, the Engineering Manager and the Budget Controller must also keep the Project Controls Manager fully informed about changes to the Project Scope, Project Cost and Project Schedule, as must the Procurement Manager, the Construction Manager and the Commissioning Manager. Monitoring and controlling those three Project Baselines is a continual team effort if the required profit is going to be realised in full.

The Board of Directors, in conjunction with the Corporate Management Team, appoints the PM to manage the day-to-day activities in respect of the Project. The PM's duty is to report faithfully to the Project Director (PD) if any problems arise that threaten the Contractor's achievement of its objective to complete the Project on time, within budget and to the required quality standard. The PM is therefore, without doubt, the single most important person to look to when it comes to whether the Project is going to be successful (and make at least the profit anticipated by the BOD) or whether the Project will be doomed to failure. A strong and effective PM can often offset the negative aspects of a weak PD to a certain extent, but even a first-class PD would probably not be able to save a Project if the PM is too weak or ineffective.

Extension of time claims due to delay brought about by scope of work enlargement (through changes considered by the Contractor to be caused from the Employer's side) account for the greatest portion of construction Disputes that end up in litigation.[9] This applies no matter where in the world one looks.[10] If the PM deliberately conceals bad news from the PD (such as the true picture about a significant delay problem), then I consider it would amount to a gross failure of his/her duties and be nothing less than a betrayal of the trust placed in him/her by the BOD, the PD and the Corporate Management Team. This is an important point, primarily because an extension of time claim can be a very sensitive issue that could cause a public relations problem, not only for the Contractor but also for the Employer. Added to that, most contractual provisions impose strict time bars for late claims. If the Corporate Management Team and the BOD are not informed about what is happening as early on as possible, it could prove too late to mend the bridges thereafter, especially if the news has already been picked up by the media.

It should be clearly understood therefore that the decision to raise an extension of time claim is not the prerogative of the PM but that of the PD, whose responsibility it is to recommend the most appropriate action to the Corporate Management Team and the BOD. Of course, under normal circumstances, the PD's recommendation would need to be based on sound advice given by the PM as to the actual status/progress of the Project and the honest reasons for any delays experienced. An abnormal situation would be where the PM and PD are in dispute about the Project's true status and the reasons behind any problems that have arisen. No matter if there are differences of opinion amongst the Contractor's Project Management Team, for the reasons given above I consider that timely flagging up of a serious delay (or potentially serious delay) problem to upper management is a fundamental obligation of the PM. However, I have been involved with Projects where this advice was not heeded, and where it then resulted in considerable financial harm for the Contractor.

One of the most vital functions of the PM at the outset of the Project is to ascertain the manpower resourcing requirements in conjunction with the PD. That exercise includes determining the start dates required for all key positions and establishing the required level of expertise and qualifications of potential candidates. This cannot possibly be done properly or effectively without a valid Project Baseline Schedule having been prepared first by the Project Controls Department, to a sufficient level of detail that will enable meaningful manpower and construction resourcing histograms to be prepared. Although such a requirement may seem obvious, I have worked on a number of mega

9 Tolson, S. (July 2013). *Disputes in the construction industry*. https://www.fenwickelliott.com/sites/default/files/simon_tolson_-_qanda_for_lawyer_monthly.pdf (accessed 18 August 2017).
10 Almutairi, S. (August 2015). *Causes of litigation in the Saudi Arabian construction industry*. https://repository.asu.edu/attachments/157972/content/Almutairi_asu_0010N_15165.pdf (accessed 13 April 2018).

Projects where that requirement was not fulfilled, and where the final number of man-hours exceeded the original estimate considerably. On one occasion that resulted in the final man-hours total being more than three times the original allowance for, essentially, an unchanged Scope of Work. It does not take much imagination to see what happened to that particular Project's profitability when it is considered that the originally anticipated manpower bill represented about 10% of the Contract Price and that the originally anticipated profit level was only 7.5% of the Contract Price.

As mentioned above, the PM reports directly to the PD, who in turn ultimately reports to both the Corporate Management Team and the BOD. To avoid confusion and the possibility of receiving conflicting instructions from on high, this arrangement should be made clear to all the Project's participants, in order that there is no confusion as to the correct chain of command. It would be a recipe for disaster if any Manager is placed in the position of having to take conflicting instructions from two people higher up the corporate ladder. It would of course be perfectly permissible for the PD to stand in fulltime for the PM whenever the PM is on leave. The taking up of that temporary position should be communicated to everyone in advance, and then relinquished by the PD immediately upon the return to work of the PM. As obvious as the foregoing observations may appear to be, I have witnessed situations where Managers have been given instructions by the PD that differ from those received from the PM. That resulted in a certain degree of chaos, as well as damaged egos and frayed tempers.

6.4.2 PM's Reporting Duties

The PM reports to the PD, who is higher in the company's hierarchy and, ultimately, will be held responsible for ensuring that the Project is a success. This is because the PD will have been given the authority and power by the BOD to make all decisions regarding the Project's implementation activities. However, the PD may be required to oversee several different Projects concurrently, which could mean that his/her focus may be distracted at a critical time for the PM's particular Project, according to what is happening on those other Projects. It is therefore vital that any unusual/exceptional issues are brought to the attention of the PD as quickly as possible by the PM and not left in limbo for too long, simply waiting for the PD to catch up with events.

The PM is charged with utilising the talents of the Department Managers to the Project's best advantage. In order to understand how the Department Managers are functioning (be that well, middling or poorly), the PM must be provided with data that measures the performance of the various Departments in meaningful ways. This must be done by the PCM setting up agreed SMART-based[11] Key Performance Indicators (KPIs) jointly with the PM and the relevant Department Manager. The PM's greatest risk arises where the Departments are not functioning well in terms of progress or in meeting quality, heath, safety and environmental requirements, and this is what the KPIs must track as a priority. Each week the PM must sit down with the entire Project Management Team and review the status of the work of each Department, using the KPIs produced by the

---

11 SMART for KPIs = Specific, Measurable, Achievable, Relevant and Time-bound. Credit for the term is given to George Doran G., Miller A. and Cunningham J. and appeared in the November 1981 issue of *Management Review*.

Project Controls Department. Where the KPIs indicate that there are problems, then the PM will be responsible for determining the best course of action, and whether or not the issues go beyond the level of the Project itself and should therefore be brought to the attention of the Corporate Management Team and the BOD via the PD. Sadly, I have seen this done far less often than it should have been, which meant that there were many occasions where problems were discovered far later than should otherwise have been the case.

What the PM must not do is get involved in directly undertaking the work of the Department Managers. If the PM ever feels the need to do this, then that is probably the signal that it is time to replace an underperforming Department Manager. Failure to do so will inevitably sap the PM's energy, confuse the relevant Department's staff and distract the PM from dealing adequately with those matters that are properly the sole responsibility of the PM. It would be far better all round if the PM were to give clear direction as to how a Department's problems should be sorted out and, thereafter, monitor the situation closely to ensure that the concerned Department Manager is then coping satisfactorily. If the Department Manager is still not performing as required, it would then be preferable for the PM to replace him/her with somebody more suitable.

Both the PM and the PD are very dependent on the quality of the reports produced by the various Departments involved, in terms of comprehensiveness and veracity. The PM should not expect the PD to pick up on errors and inconsistencies in the reports. Instead, mechanisms should be put in place that act as double-checks on whatever is being reported, since it is a fact of life in the construction industry that Managers constantly overstate the value/quantity of work achieved by their respective Departments. For example, the Engineering Manager may claim that Detailed Design is 90% complete, whereas an independent check on the completeness of the drawings would reveal that there are a significant number of major 'hold' points on 15% of those drawings (meaning that the level of completeness may be closer to 80% than 90%). Another problem area is where a large number of Vendor Documents are late in being submitted to the Engineering Department, yet it is claimed that, overall, such documents are 99% complete. These misreporting incidents can severely mislead the Contractor into being over-optimistic about the time needed to complete the remaining construction work and achieve Mechanical Completion on the Project. The PM needs to be continually aware of this tendency for reports to conceal where delays are sitting, and bear in mind that it is essential to find alternative ways to measure the level of progress being reported for all the various Disciplines and Departments (to act as a checking mechanism).

Another area where the PM is reliant upon honest reporting is in regard to the quality of the ongoing construction work. The Employer is paying for a completed Project that, in terms of functionality, operability, maintainability, quality and safety, meets with all the requirements that were set out in the Contract Documents. And that is precisely what the Employer expects to get and is entitled to receive. However, all too often the list of Punch Items that need to be rectified before commissioning work can commence is crippling, and the true 'quality' picture only becomes apparent too late in the day. The PM must therefore put mechanisms in place to ensure that this situation of an over-extensive list of Punch Items is not allowed to occur. Such a situation usually arises because too much emphasis has been paid to getting the job finished as quickly (and cheaply) as possible, which then inevitably led to substandard workmanship.

The PM also needs to ensure that what develops with the Employer's Team is faithfully recorded too. On many a Project, the Employer's representative insists on producing the Minutes of Meetings (MOMs), which are then compiled to the advantage of the Employer. It is incumbent on the PM to ensure that all MOMs accurately reflect what was discussed and agreed upon between the parties, especially: (i) the decisions that were reached, (ii) the actions that were decided, and (iii) the progress status. If the PM does not make it his/her personal duty to read all proposed MOMs before signature, then inevitably some items will be recorded incorrectly that may then cause later problems for the Contractor. Consequently, where it is observed by the Contractor's Team that the Employer's representative has misquoted something in the MOMs, especially with regard to progress concerns, then such matter should be addressed immediately, with a request to have the MOMs corrected.

If the Employer's representative refuses to amend any offending MOMs, then the PM should refrain from signing them and, instead, formally write to the Employer stating why the Contractor objects to the MOMs as tabled by the Employer's representative. Producing contemporaneous notes in a court of law that are at odds with the contents of the signed MOMs will not achieve anything worthwhile for the Contractor. The best way I saw MOMs being handled was agreeing to the wording of the individual agenda items as the meeting went along, and then getting everybody to sign off before leaving the meeting. This enabled all disagreements about what was said (or meant) to be sorted out there and then.

6.4.3 Preparation of Project Execution Plan

At the commencement of any major Project, the Contractor will be required to prepare and submit the finalised Project Execution Plan (PEP), setting out the Contractor's plans for managing the Project, from mobilisation right through to close-out. The responsibility for getting this organised falls to the PM. If prepared properly, the PEP can be a very useful document for people to refer to quickly in order to see what the whole Project is about, as well as appreciate what the specific roles and responsibilities of all the various participants involved in the Project are. All Managers must be made to be directly responsible for inputting into the PEP the details relevant to their own Department, including the specific scope of work that their staff will be required to undertake. It would be a big mistake for the PM to free up the Managers to work on other issues and not insist on them individually giving their best input to the PEP as their first task. The whole idea of the PEP is that it becomes the principal guide to everybody as to how the Project will be managed. All the Managers should therefore be made to contribute to the PEP properly and assume ownership of the content as far as it relates to their respective Department's work. As and when necessary, such as when the implementation strategy changes, the PEP should be updated and the new version then circulated to all the Managers as a matter of priority.

I have on many occasions seen Invitation to Bid documents and the subsequent Contract Documents calling for the Contractor to prepare unnecessary documentation of various sorts. Contrary to what some Employers seem to insist upon, I personally advise not wasting too much time in preparing a detailed Executive Summary for the PEP. As far as I am concerned, calling for an Executive Summary to be inserted at the start of the PEP documentation offers absolutely no benefits whatsoever and can even be considered to be detrimental. As mentioned earlier, the PEP is supposed to be a dynamic document, updated at regular intervals to reflect any significant changes that have occurred in the Contractor's implementation strategy. Having to modify an over-detailed Executive Summary every time an update was produced was the excuse I was given as to why the PEP update was continually being delayed. Those delays rendered the PEP less useful than it could have been. It took a great deal of effort to produce the original of the PEPs I saw semi-abandoned in that manner, and not being able to make full use of the PEP to help drive the Project through to completion more effectively was a great pity in my judgement.

If the provision of an Executive Summary is really insisted upon by the Employer's Team, my advice is to keep it to the bare minimum. My reasoning for that is simple. It is because the proper function of an Executive Summary is to summarise the contents of an analytical study. That then allows the reader to see both the conclusion and the main pointers as to how the conclusion was reached, without the need to read the whole study. Ordinarily, an Executive Summary should be written in such a way as to make it possible for the reader to decide which sections of the main body of the work need to be inspected in more detail. This is important where the confidence level of the conclusions reached in the main body of a document needs to be determined in the reader's mind. However, the PEP does not have a conclusion to report on; it is, in effect, an instruction manual. Reading many of the Executive Summaries of the PEPs I have seen was, I felt, a complete waste of time. Many of them were little more than the main body of the PEP in bulleted form, taking up plenty of space but offering nothing worth reading. Further, where attempts had been made to précis the main PEP content, I saw that it sometimes provided an excuse for lazy people not to bother to read the main body of the PEP.

If an Executive Summary for the PEP has to be included, my advice is therefore to produce an outline Executive Summary only, with such little detail that it forces people to read the main content of the PEP. The Contractor should first concentrate on writing the PEP up in a structured, story-telling fashion, making sure that the document is divided up into useful bite-sized chunks. Plenty of appropriate headings and sub-headings need to be added, together with a contents list in the front, so that people can quickly find what it is they need to see. In this way, the Contractor will have constructed a truly useful document that can be referred to by many on the Project when undertaking the wide variety of important tasks that are necessary on every Project. It would also be helpful as a tool for quickly identifying the Project's major risks. And it will have obviated the need for providing an Executive Summary in any form.

6.4.4 PM's Responsibility for Risk Management

Much is made today of the need to engage a full-time professional Risk Manager for managing the risks for major EPC Projects, but I contend that this is generally unnecessary except for very complicated Projects. What needs to be done instead is to appoint a PM who has demonstrated on previous Projects that he/she understands the importance of identifying and managing a construction Project's risks. This will require the proposed PM showing that he/she has managed his/her previous Projects successfully in the situation where he/she also undertook the role of Risk Manager at the same time. Specifically, the proposed PM should be able to show that the Scope, Cost and Schedule Baselines were properly controlled, and that adequate reimbursement and/or appropriate extensions of time were obtained on those previous Projects for all authorised deviations from those Baselines.

It should be borne in mind that most of the issues that have the tendency to negatively impact performance on a Project will result, in one way or another, from improperly managed risks. A good PM will attempt to lessen the negative impact of risks significantly by (i) implementing a policy of open communication with the other Managers, and (ii) ensuring that all key participants are given the opportunity to express their opinions and concerns without fear of being reprimanded. Employing such a policy offers the best route for the PM to be notified of new potential risks early on, so that any problems observed can be nipped in the bud quickly, before any major damage occurs.

If the PM's primary areas for implementing risk management were to be encapsulated in a list, I suggest that it should read something along the following lines:

- Identifying and managing the Project's major risks;
- Helping to develop the Project Execution Plan;
- Organising, monitoring and controlling the Project's resource needs;
- Ensuring safe working at all times;
- Directing and monitoring the Project's management staff;
- Monitoring and reporting on the Project's actual progress;
- Developing/implementing appropriate plans to redress delays;
- Completing within the required time-frame;
- Delivering the expected quality of the end product; and
- Completing within the approved budget.

6.4.5 PM's Key Considerations for Minimising Risks

The above list may need to be re-read a few times to understand the full implication of the extent of the responsibilities it contains for the PM, because there are huge risks attached to achieving each of the identified functions/activities satisfactorily. In order to minimise those risks as much as possible, the PM therefore needs to take the following into account when dealing with the carrying out of the respective function/activity:

6.4.5.1   *Identifying and Managing the Project's Major Risks*
Unless the BOD considers that a specialist Risk Manager needs to be brought on board due to the unusual or onerous nature of the Project, the PM must be the one to take the lead role in this activity. However, it must be undertaken as a joint effort from all other Managers inputting their ideas. That must include even such people as the

Commissioning Manager, regardless of the fact that later events might change the course of the Project further down the line before such personnel are needed on the scene fulltime. If the PM undertakes this task in isolation, then there is a likelihood that the other Managers will not attempt to assume ownership of the risks that impact their own Department's work. I have observed first-hand that such situation can prove disastrous for the well-being of the Project.

6.4.5.2   Helping to Develop the Project Execution Plan

The overall strategy for the Project should be discussed fully and agreed amongst all members of the Project Management Team (as mentioned above), with the objective of it forming the basis of the content of the PEP. Following such discussions, the individual Managers should, for the PM's review and agreement, each write the details of precisely how the work of their respective Departments will be organised and how interfaces with the work of all other Departments will be managed. The primary focus should be, as much as possible, to properly identify and offer meaningful mitigation measures to counter the major potential risks to the Project. If this is not done properly and no clear plan exists, there will inevitably be many times when things will go wrong unexpectedly that could then jeopardise the success of the Project. The PEP is a vital tool that needs people committed to producing it and following it. The risks involved in not doing that include nobody having a clear, overall picture of what is going on. If Managers do not commit to writing a meaningful PEP, then it will mean that the plan is only half thought through and thus will be far less useful than it could have been; further, there is a good chance of such a poor plan going wrong.

6.4.5.3   Organising, Monitoring and Controlling the Project's Resources

The PM should be the only one to make the final decisions as to what elements of the work are to be sublet, whether construction equipment should be hired or bought outright, what the overall manpower resource levels on the Site must be, and so on. If others are allowed to do this then, as a minimum, cost control will become impossible. However, when it comes to the staffing plans necessary for the individual Departments, it seems quite obvious that these need to be discussed and agreed with each respective Manager first. Nonetheless, I have been amazed to witness a PM unilaterally deciding that the Procurement Department did not need Expediters to follow up on Vendor Documents or chase up on Vendor deliveries. The expectation had been that, somehow, the Buyers could quite easily cope with those extra tasks; instead, I witnessed very clearly that they failed to do so adequately. The failure of the goods to arrive on time under those circumstances cannot be laid on the doorstep of the Procurement Department, except in the situation where the Procurement Manager failed to raise the negative impacts of the understaffing problem with the PM.

I have seen decisions such as that described above taken by PMs for the purpose of keeping costs down, presumably in the hope of securing their end-of-Project bonuses. However, the fact remained that the workload was unfairly increased for others who did not derive the same bonus benefits but who had to put up with considerable extra pressure to get their work done on time. This is therefore a good reason why the PM must be charged with obtaining the inputs from both the Project Controls Manager and the Manager of the relevant Department to the required manpower resourcing levels, and for the Project Director to be responsible for endorsing those findings following separate discussion with those others. It is entirely unreasonable for a Department Manager to be held responsible for failing to meet deadlines with inadequate staffing numbers that he/she never agreed to and quite clearly objected to.

6.4.5.4 *Ensuring Safe Working at All Times*

The welfare of all the Contractor's workers, as well as that of all others visiting or working on the Site, must be a paramount consideration of the PM. This is not least because, in many countries (such as the Philippines, for example), the PM will be arrested pending investigation if a serious accident occurs on the Site. The sad truth is that, despite all the training given nowadays that instructs workers to do otherwise, many of them are still apt to take unnecessary chances in the workplace. Such actions put not only their own lives in danger but sometimes the lives of their co-workers too. Notwithstanding that, it is not the PM's job to monitor the workers to ensure that they are working safely. However, it is the PM's job to ensure that sufficient numbers of suitably qualified staff (especially Health, Safety, and Environment [HSE] personnel) are available at all times, both to provide adequate safety training to the workers before each critical activity is undertaken and to monitor closely the implementation activities that follow. This requires specific 'Toolbox Talks' to be given (with special training provided where necessary) to highlight all the key essentials to be taken account of when undertaking the physical work. If such instructional talks and necessary training are not given, it will increase the chances of the work not being completed without something untoward happening and workers possibly being injured (or worse).

Different Employers have different ideas as to the ratio of HSE Supervisors to workers that needs to be provided on a construction Site. A general consensus of opinion seems to be that an absolute minimum of one HSE Supervisor should be provided for every 50 workers, even on a standard, straightforward Project. If the work is especially complex and/or dangerous, then a ratio nearer 1:30 may be considered more appropriate. Since HSE personnel are no different from other workers (they get sick occasionally and go on vacation too), the PM must make sure that adequate cover is available to cope with those absentee situations.

Most 'Conditions of Subcontract' make the Subcontractors responsible for providing their own HSE personnel, but that will not necessarily absolve the PM from responsibility if something disastrous happens that is caused by a Subcontractor's own personnel. It is therefore essential that the PM makes arrangements for one of the Contractor's HSE personnel to attend the Toolbox Training sessions undertaken by each of the Subcontractors. That same person must also be made responsible for ensuring that all Personal Protective Equipment is available for all the workers who will be involved in undertaking the physical work. A good control mechanism here for the PM's benefit is to see that Toolbox Training is included in the list of KPIs, and that all participants of each training session sign off the official record of the training (which should be lodged each day for filing by members of the Document Control Department).

The PM must not be lax when it comes to disciplining workers responsible for violating the safety rules, since it would send the wrong signals to the other workers, who might then be tempted to ease off employing the rules to the full extent required. The PM must also be vigilant to the HSE personnel slacking in their monitoring duties. One way around that is to insist that, before the start of each new day's work, all safety incidents that occurred during the previous day should have already been reported to the PM in writing by each individual HSE Supervisor. Such incidents should then be discussed frankly with the PM to see what lessons can be learnt. Again, all such reports should be lodged daily for official filing.

Many accidents on the Site occur due to the workers trying to get things done quickly, sometimes before all safety checks have been conducted, especially for such activities as the lifting of heavy goods/equipment. The PM should occasionally swoop on such activities in progress, and ask to see the records of the safety checks that were conducted. If it is discovered that unacceptable shortcuts were taken, then the PM should make a big fuss so that the workers are left in no doubt that he/she does not require, and will not tolerate, such shortcuts being taken.

6.4.5.5 Directing and Monitoring the Project's Management Staff

No one person is capable of doing everybody else's job. If a PM is incapable of recognising the importance of the contribution of his/her Managers and does not employ their talents effectively, then the Project will almost certainly be doomed to failure. Some PMs I have had the misfortune to meet should never have been placed in that important position, because it soon became obvious to the whole team that they did not know how to manage people. With a couple of notable exceptions, those PMs who could not manage people properly were very capable technically, which is most likely why they were given the position of PM in the first place. But that is where the mistake was made by the Corporate Management Team (or, more specifically, the Chief Executive Officer). To reduce the risk of this situation materialising, my advice is that anybody proposed for taking on the role of PM for a major Project must be well known to the Contractor and, preferably, already tried and tested under the direction of one of the company's trusted Senior Managers before being let loose. If that advice is ignored, then the Chief Executive Officer will have to assume responsibility if devastating personality clashes occur inside the Contractor's Project Implementation Team and/or with the Employer's Team.

In reality, with all other things being equal (such as the Contract Price and the Project Schedule being reasonable), the success of an EPC Project rests more than anything else on the ability of the PM to:

(i) assess the strengths and weaknesses of the members of the Project Implementation Team and take appropriate steps to ensure that all capability gaps are closed,

(ii) provide adequate support/training wherever capability weaknesses are evidenced (with the help of and through the efforts of the Corporate Management Team, of course),

(iii) utilise to the full the strengths of each and every person (encouraging them to maximise their contributions), and

(iv) ensure that adequate, genuine team-building activities are conducted, that everybody is able to stand together behind the creed of 'One Team, One Objective', and that everything possible is being done not only to retain key personnel but to earn their loyalty too.

Where many Project Management Team members fall down is in failing to report faithfully when and where things are going wrong. Instead, a Manager will often try solving a problem single-handedly and then only expose it too late in the day, after the situation has become toxic. To counter this, the PM needs to ensure that an environment of 'no blame' is cultivated on the Project. This will have the effect of emboldening all Managers to speak up as soon as a problem surfaces, giving the whole Project Management Team the opportunity to resolve the problem in good time, before it has a chance to get out of hand.

The most helpful way for the PM to know what the Department and Discipline Managers are doing, and how they are progressing, is to insist on receiving an updated weekly 'To Do' list from each of them, in which they identify a list of their prioritised activities for the week. On subsequent weeks, the previous week's prioritised list should still be attached but updated, in order to indicate what was actually achieved in the previous week. This will prove very useful if a Manager suddenly becomes unavailable due to illness or because emergency leave becomes necessary, since whoever is then required to take over can see exactly what remains to be done.

In regard to that last point, a mistake is often made by the PM letting somebody depart on leave without having first provided adequate handover notes. Those notes should not be limited to just the current and next week's activities but should cover a full month of activities from the intended day of departure. They should also be complete with a copy of all associated backup paperwork relevant to the matter. I have heard all sorts of excuses made by Managers for not providing handover notes, but none of them has ever held water in my opinion. Pressure of work is no excuse at all and, if the weekly 'To Do' list has been faithfully kept and updated, it ought to provide a good basis for the handover notes for both the current and following week's activities. Adding on details about the final two weeks' anticipated activities should not prove difficult for a competent Manager. My advice is simple: nobody should be allowed to go on leave until his/her handover notes have been satisfactorily submitted by close of business one day before the last full day of working.

In addition, it is essential to obtain the personal telephone number of the departing Manager on which he/she can be contacted at all times while on leave, in case an urgent matter arises that requires his/her personal input. Insisting on this whenever key personnel are going to be absent from the Site will help reduce the risks that essential matters will be forgotten until too late and therefore not dealt with in good time.

6.4.5.6 Monitoring and Reporting on the Project's Actual Progress

The PM is responsible for officially reporting the Project's progress in two directions: (i) to the Employer's Team and (ii) through the Project Director, to both the Corporate Management Team and the BOD. Such reporting must be exactly the same for all recipients, and be presented so that it can be seen clearly and quickly if the Project is doing well or veering off track. The principle issue the Employer will pay attention to is the Project's progress status, and most Contractor's focus on that in their reporting, along with the HSE statistics. On the other hand, I have observed that the average Contractor's reporting is generally very light when it comes to cash flow matters, and virtually non-existent when it comes to the issue of income versus outgoings. That would perhaps not matter too much if those issues were being fully and openly discussed up to the level of the Contractor's Corporate Management Team. However, I have observed that it is not unusual for PMs to sit on financial bad news until the problem has developed into a cause for major concern.

Cash flow is vitally important to every contracting organisation, since lack of cash in the bank can result in the Contractor's whole payment system being crippled (not just for the Project but for the entire business). If ready cash for vital payments becomes scarce, it can then lead to late material supplies and inadequate labour resourcing problems for the Contractor. In an effort to mask looming financial problems for as long as possible, I have observed some PMs push the Cost Management Team to inflate the progress measurement figures in order to increase the amount received for interim progress payments (over-claiming). Such a tactic can cause a number of other problems later, as the following demonstrates:

1. Access to the progress reporting output is not limited simply to the teams responsible for handling progress payments but is also fed into the weekly reports that are supplied to the Employer. If a situation occurs where, later, an instruction from the Employer begins to delay progress of an activity where progress has been over-claimed, any extension of time claim will have to be developed from the starting point of the last recorded progress situation for that activity immediately prior to the delay event occurring. This is because, legitimately, the delay can only be shown to impact the work that still remains to be done (based on a reading of the previously approved progress report). This can severely compromise the Contractor when attempting to obtain a legitimate extension of time.

2. Especially where the over-claiming is carried out on a wide-scale basis on a large Project, those responsible for arranging follow-on activities may be fooled into planning for the subsequent work to be commenced a lot sooner than is actually required. This can lead to frustration for, and even claims from, affected third parties, and that is not good for overall morale on the Project.

3. If certain people are not aware of the over-claiming and rely upon the gross progress payment figure to guide them as to the status of the Project's progress, then uncalled for complacency can set in, leading to loss of productivity that could have dire consequences for the Project.

4. The BOD may be misled into believing that the profit that seems to be generated by the Project is real, when in fact there is no profit at all. Even worse, the funds perceived as being surplus may not even pay for completing the work that has been over-claimed, let alone produce a profit. If the BOD then mistakenly reduces external funding arrangements or takes on additional large financial commitments in the mistaken expectation of substantial profit arising from the Project, that could have disastrous consequences for the Contractor's entire operations.

If it is discovered that over-claiming of work progress has been done (whether accidentally or deliberately), the most important thing for the PM to take care of is to ensure that the true Project progress and associated financial repercussions are reported to and fully understood by the PD, the Corporate Management Team and the BOD. Leaving the nasty surprises until later will not be appreciated by anybody. Also, I observed that knowledge about such over-claiming matters had a horrible habit of being leaked and reaching the Employer's ears by indirect routes. That then had the effect of severely denting the Contractor's credibility with the Employer. It is therefore better for the Contractor to come clean with the Employer about such over-claiming rather than waiting for auditors to pounce upon the problem. That is especially true in today's climate, where compliance with the Sarbanes-Oxley Act[12] is a must for most companies. I have witnessed such honesty and transparency from the Contractor rewarded by the Employer agreeing to recover the overpayment incrementally over time instead of insisting upon immediate repayment of the overpaid amount. I have also seen examples of the Employer demanding immediate repayment where the Employer's Team had unearthed an overpayment problem that the Contractor had been attempting to keep covered up.

The ultimate goal of the PM is to get the Project completed on time and within the budget allowance (all without sacrificing either worker safety or the Employer's quality objectives, of course). To be able to understand not just the true progress status of the Project but also the true financial status, I consider that it is essential for a full Earned Value Management System (EVMS) to be put in place and properly operated by competent, experienced personnel.

The PM must take the time to learn how to interpret the outputs from the EVMS that will be produced by the Cost Management Team (operating as part of the Finance Department). This is because an EVMS helps considerably in providing an answer to what is often a much undervalued concept in progress measurement. That concept is to measure, at any given stage of a Project, exactly what is needed to get the Project completed in terms of time, money and, ultimately, resources. For example, if, for a given activity, the income to date is 70% of the total payment due for that activity but only 35% of the work has been completed, then it should be obvious that the balance of 30% payment that still remains will be inadequate to cover the costs for the remaining 65% of work that has still to be done. This is where operating an EVMS could save the day by helping to identify such problems as these before it becomes too late to be able to do anything about them. However, the application of EVMS is not as well understood by as many Managers as perhaps it should be, even though its efficacy has been adequately demonstrated.[13]

Of course, to be able to get all the necessary progress and cost reports, the Project Implementation Team must in the first place have sufficient numbers of staff in the Planning and Cost Management Departments; it is heavy work, and an under-staffed team will not cope adequately. The PM must then fully utilise the efforts of the progress measurement team to ensure that he/she is kept fully informed about the Project's progress via submission of properly substantiated reports. Those reports should not be

---

12 The USA's Sarbanes-Oxley Act (2002), named after Senator Paul Sarbanes and Representative Michael Oxley (the principal architects of the legislation), which introduced major changes in the regulation of financial practice and corporate governance.
13 Association for Project Management (2013). *Earned Value Management Handbook*. See the Foreword regarding EVMS successes in Great Britain for the 2012 Olympics.

accepted at face value but should be scrutinised in detail by the PM, in order to reduce the risk that the Project's progress is not being monitored or reported properly.

Ultimately, the PM needs to work closely with the Planning Team in order to be assured that what is being shown as Project progress is realistic, and that catch-up plans are valid. I have worked with PMs who have ignored the outputs from the Planning Team, and even a few who were not concerned when a Senior Project Planner was never appointed or resigned part-way through the Project, leaving only junior personnel to present progress as 'history', without considering what lies ahead. On quite a number of occasions it was later necessary to call in third-party specialists to retrospectively create the progress charts and prove that the Contractor was entitled to an extension of time. All of that cost a lot of additional money and took a great deal of time, during which period the PMs concerned had to sweat it out to see if they still had some credibility left with their respective Boards of Directors. Some of them found that their reputation was nowhere near as good as it had been previously.

In a nutshell, failure of a PM to focus on keeping the Project moving forward or simply leaving others to get the Project completed on time is a risk that no competent PM should ever take. In a healthy economic environment, the Employer will always want the Project completed on time, and will get very upset when completion is delayed not due to the Employer's doing. That is where Liquidated Damages (LDs) come into play that could see the PM's future job prospects with his/her current employer liquidated too. The PM should take steps to remind him/herself every day that the value of full LDs on most construction Projects is often greater than the Contractor's anticipated profit for those Projects.

6.4.5.7 Developing/Implementing Appropriate Plans to Redress Delays

In my book, time is the most critical resource on any Project, and no efforts should ever be spared to ensure that it is used effectively. Failure to complete the Project on time due to the fault of the Contractor will negatively impact the Contractor's reputation and cause concerns for future prospective Employers (meaning that future bidding opportunities may be reduced). Added to that, there is the very real risk of the Employer demanding that the Contractor must pay Liquidated Damages for delayed completion. Further, the Contractor will inevitably incur additional management and administration costs due to the construction period being extended.

The clock starts running for the Contractor from the moment the commencement date stipulated in the Contract kicks in, yet I have sometimes witnessed very little or no action being taken for several weeks after the Project's time-line has commenced running. This situation is almost unforgiveable, since most EPC Projects will require a substantial amount of procedural documentation to be submitted for review within the first 30 days or so of the commencement date. That requirement will have been known about well in advance of the Contract signing date. It is therefore incumbent on the PM to ensure that the appropriate teams, with the right numbers of personnel on board, are ready and available to commence working on all the initial activities from day one.

Ideally, the Project must be seen to get off to a flying start in order to cultivate the right atmosphere amongst all members of the Contractor's Project Implementation Team. The alternative approach (i.e. letting things drift a little at the outset) will set the wrong tone completely, and it can then sometimes take weeks to rectify the opinion of the general staff that the Contractor's Managers are not concerned about time, and that there is no urgency attached to the Project's implementation. However, nothing could be further from the truth. Time is money and, once time has been lost on the Critical Path through problems that are caused from the Contractor's side, it will always be very difficult to recover from such delays; recovery of the additional expenditure involved will be impossible.

The PM's responsibility is to ensure that every single Manager on the Project not only understands how important adherence to the Project Schedule is but that they also convey that message to every single individual within each Manager's respective Department. Ideally, those messages should be conveyed formally and go out before the Contract's commencement date, not after, by way of small team meetings chaired by the various Managers. The intent must be to ensure that everybody understands that programme slippages from the Contractor's side must be rectified immediately and not allowed to mount up. In this way, the individual Managers will each be made responsible for adhering to the Project's time-line. The Managers should also be charged with the responsibility of reporting back to the PM as soon as it becomes apparent that keeping to the schedule in his/her particular Department will not be possible using only the currently available resources. As soon as a delay problem is notified to the PM, the Managers of all the Departments that may be affected by that delay should be called to a meeting so that all major issues surrounding the delay situation can be fully discussed. All valid proposed remedies can then be tabled, and a workable solution for rectifying the problem should be agreed upon as quickly as possible.

Instituting overtime working is sometimes not a viable solution, as it can exhaust the workers and lead both to decreased productivity and shoddy workmanship (as well as to sickness if stress has set in). Whatever the solution is to overcoming a delay problem, the result needs to be monitored closely to see that it has the desired effect of allowing progress to catch up with the original programme. If it is not doing that, then a rapid rethink must be carried out amongst all the affected Managers, headed by the PM, so that a truly workable solution can be found.

6.4.5.8 Completing within the Agreed Time-Frame

My basic training was as a Quantity Surveyor, and I have never forgotten the fundamental principles of that honourable discipline. I take its lessons with me into every new Project I become involved with. As a Quantity Surveyor, my foremost consideration was always with the costs and revenues associated with a Project's implementation work. With my Contractor's Quantity Surveying hat on nowadays, I still wish to see a healthy profit coming out of all the hard work that goes into completing an EPC Project. However, many years ago, long before I became involved in EPC work, I observed a phenomenon that has never varied. That was the simple fact that Projects where time was controlled well did better financially than those where time was out of control. As far as I am concerned, it is therefore of paramount importance that the PM should concentrate on everybody and every element of work adhering to the Project Schedule.

I have heard people say words to the effect that the Project Schedule is an indicative guide only as to what will happen and should not be taken too seriously. As a direct result of my observations and experience regarding management of time, I have an entirely different take on the matter. For me, the Project Schedule must be considered to be written in stone until such time as it is necessary to modify it to fit changed (or changing) circumstances. This is because it is essential that all the Managers buy into the Project Schedule and strive to achieve it. If just one Manager responsible for critical activities decides that the Project Schedule can be ignored, then the delayed work from his/her Department will inevitably have a negative knock-on effect on the work of all other Departments that rely on the outputs from his/her Department. The PM must ensure that this is not allowed to happen.

There is only one way to get every Manager to accept that the Project Schedule is sacred, and that is to get each and every one of them fully involved, and make them responsible for the inputs to the Project Schedule in respect of their individual Departments. Sadly, I have rarely seen this done. And that probably accounts for why most EPC Projects I have been involved with ran into serious critical delays. In fact, many of the Managers I spoke to (when trying to put together a defence for the delays), expressed dissatisfaction with the tight time-frames imposed on their respective Departments and the limited resources they were given with which to complete their work.

The problem with the inputs to the Project Schedule usually start as far back as the bidding stage for the Project. The first hurdle to clear is often the ridiculously short time-frame in which the Employer expects the Project to be completed. More often than not, the Contractor has to live with the tight completion date required. The second obstacle is the costings, which need to be kept as competitive as possible in order for the Contractor to stay in the frame for Contract award. This generally means that the manpower and equipment resources have to be pruned down as much as possible.

Unfortunately, the Contractor's pricing problems may not end there. This is because many Employers will still demand a discount, even from the lowest bidder, which reduces even further the amount of money the Contractor will have for providing additional resources. This is despite everybody in the business understanding that fair, open and transparent competitive bidding means that the lowest technically acceptable bidder should be awarded the job of undertaking the Project's implementation work, without further reducing the original bid price. The bidders would be severely punished if they were found to be colluding on the bid pricing, but many Employers have no qualms about flouting the fundamental principles of fair competitive bidding. What all this adds up to is that many Contractors commence an EPC Project with much less money in it for them than had been hoped for, which means that the resources necessary to undertake the Project have to be very carefully controlled. It seems that scant attention is paid by too many Employers to the argument that even the un-reduced original competitive bid pricing route should be avoided (let alone demanding a reduction thereto).[14]

---

14 Odgers, K., Rowsell, S., Thomas, K. et al (updated April 2011). *The business case for lowest price*

Despite all the monetary pressures, the PM must nonetheless bring all other Managers on board when it comes to agreeing collectively to adhering to a Project Schedule that has been pre-compiled and is desperately shorter than required. The question therefore remains: how best to do that under the scenario described above? As I see it, the following needs to be done immediately after award of the Project.

1. Each Manager must first draw up a detailed list of everything that needs to be done in his/her Department and then allocate adequate time and personnel resources against each item in the list.

2. The PM must individually discuss the resourcing outputs from each Manager, with a view to identifying where significant differences sit when comparing those outputs against both the time-line and the budget allocated to each Department.

3. In conjunction with the Project Controls Manager and the Planning Manager, the PM must decide how best to modify the Project Schedule to reflect, as much as possible, any changes required that arise from undertaking the above exercises.

Almost inevitably, there will be cases where the expectations of one or more Managers regarding additional time and/or additional resources cannot be met either partially or in full. This will present a very tricky and sensitive situation that needs to be handled by the PM in such a way as to keep the affected Manager(s) as committed as possible to managing within the time-frame allocated and the resources available. I have seen some PMs take an irrational approach to dealing with this by reducing the overall time period for the on-Site construction work, in the belief that overtime working will somehow avoid any delays caused by extending the time-frame for the pre-Site activities. All that happened, in almost every case, was that the time period remaining for commissioning was severely reduced, so much so that the Project ultimately suffered delayed completion. In my opinion, it would have been far better in all such cases if the PM had bitten the bullet at the outset and opted for providing additional resources for the front-end activities. In that way, the Project would have stood a reasonable chance of staying on programme. As it happened, the inevitable lateness of the early activities had a demoralising effect on the Project Management Team, and the Project Schedule was then disregarded by some Managers. That is the situation the PM must take all steps to avoid.

6.4.5.9 Delivering the Expected Quality of Finished Product

One horrible surprise that can happen on an overseas Project is where the majority of the specialist tradesmen/artisans employed earlier have completed their work and been demobilised, following which it is then discovered that a vast amount of remedial work is required for their specialities. This can become a nightmare for the Contractor if the Employer's Team refuses to accept such work as Punch Items that can be attended to after the Completion Certificate has been issued. This is made even worse if demobilisation involved flying those workers back to their home countries after having had to cancel their visas in the foreign country where the Site was located.

Such a situation can arise when the workers have been left on their own too frequently, without adequate numbers of Quality Control (QC) personnel having been allocated to check the standard of the ongoing work and the quality of the finished product. It can also arise if the quality checkers spend too much time in the office and not enough time out on the Site, as very often happens. Another situation where quality problems can occur is where bonuses are offered for completing the work early, and quality is two worker crews are the same, and even from one day to the next the outputs from the same crew in terms of quality can vary dramatically. Most people who have faithfully followed their favourite sports team over the course of a few seasons can probably relate well to this problem.

One way to try to reduce both of the foregoing problems is to consider if bonuses can be offered for the workers for limiting the number of Punch Items below an agreed ceiling. However, it is much harder to implement this than to talk about it, so it is not recommended as an option (although I have seen it attempted, with varying but mainly lukewarm degrees of success). I have also seen attempts made to agree lump-sum mini-contracts in advance with the workers for completing all the work, inclusive of making good all Punch Items. The problem here is that there is always a subjective element involved, where it can easily become possible for 'pecuniary advantage' to be taken of the situation, thus making the accounting difficult and, sometimes, raising suspicion that something corrupt is going on (even if that is not actually the case).

The reality is that nothing makes workers turn out better work than when they are aware that people are continually and rigorously inspecting their outputs, people who will expect the workers to put all bad work right before they leave the job. So, while it may seem that there is an added cost if more QC personnel are added to the workforce, the right balance has to be struck. One way of doing that is to consider, based on past experience, how much time could possibly/probably be saved if the QC workforce was increased by 10% (for example), and then calculate how much would be due in Liquidated Damages (LDs) if that time was not saved and the contractually required completion date was thereby missed. For example, if it was believed that adding five additional QC personnel on a 10-month construction phase would save two weeks on the completion time, the direct costs involved would be for an additional 50 man-months of input at (say) $20k all-inclusive per month (i.e. $1 million). If the LDs would be, say, $200k per day (i.e. $2.8 million for 14 days), then paying the extra $1 million out for the five additional QC personnel would seem to be a sound and worthwhile investment. At the very least, it would act as a worthwhile insurance policy against having to pay out on LDs.

Of course, nothing prevents the Contractor from building into the budget a discretionary (unannounced) bonus scheme payable where the work is good (meaning very few Punch Items are evidenced), and which can be seen to be paid out on the early work activities where good quality work has been produced. Such an approach might then encourage all follow-on trades to pay more attention to delivering quality work in order to earn a similar bonus.

6.4.5.10 Completing within the Approved Budget
When all else is said and done, the whole purpose of undertaking a Project is for the Contractor to make a profit. Finishing a Project on time is a great achievement. If timely completion is accompanied by delivering a world-class facility, then that would be another feather in the Contractor's cap. However, if all that good work is achieved but there is no profit (or, worse, a loss), then the Project will still have been a failure from the Contractor's perspective (although it is hardly likely to dent the Employer's delight at getting a beautifully crafted Project delivered on time). Sadly though, this is where most Projects go badly wrong for the EPC Contractor.

There are a number of key components that need to go into ensuring that a Project will stand a good chance of being completed within the approved budget:

- First, the budget must have been prepared adequately and be completely realistic, and it must also have been broken down into the costs allocated against each Department and each task/activity. *(Sadly, this is where a lot of Projects get off to a very bad start, and I have even been on Projects where the original budget breakdown for internal pre-bid validation purposes could not be found.)*

- Second, the Project's Budget Controller must be responsible for ensuring that each Department Manager is made fully aware of the make-up of the budget and be given the responsibility for monitoring and controlling the expenditure in his/her own Department. *(More usually, the Department Managers are kept in the dark about the budgets for their Departments.)*

- Third, all Managers must be charged with the responsibility for reporting on and managing changes to the scope of work perceived in their own Departments, in such a manner that will ensure that the current budgets for each Department are not being breached. *(I have seen that very few Managers do this satisfactorily.)*

- Fourth, each Department Manager must report back in a timely manner to the Budget Controller so that regular financial reports can be prepared and submitted to the Finance Manager for discussion with the Project Manager. *(An appropriate software package, accessible by the Managers, would obviate the need for this, but the perception that all financial matters are 'top secret' often seems to preclude this from being implemented.)*

The above principles set out what is a very transparent way to handle budget management but which, sadly, I have not seen implemented very often. The more usual system I have seen used is what I call the 'black-box' accounting system, where the Finance Department maintains utmost secrecy about all expenditure against the budget until a financial crisis has developed (usually only after the cash flow has become a very real problem). Very often the announcement of the crisis coincides with notification that the company's auditors are arriving for an impromptu visit to 'find out where things have gone wrong'.

A Project's budget problems are frequently compounded by the Managers of the Project's key components (the EPC activities) misreporting both progress and costs, very often due to the inadequacy of record keeping. In today's environment, where computers are so comparatively cheap and where so much good accounting software is available, there really should be no excuse for not being able to control the budget properly. Yet that is where many Project failures occur. As long as a Contractor's Finance Department continues to focus on recording history instead of being proactive with budget management, this situation will be repeated Project after Project. However, the BOD will generally be quick to place the blame for the financial losses on the shoulders of the Project Manager and the Project Director, even though this is the area of activity that both those individuals often have the least direct control over (since they are heavily reliant on the output from others, as mentioned earlier).

6.4.6 PM's Preferred Communication Style

I have encountered three different ways in which PMs prefer to operate:

1. Those who quite obviously do not like, or would rather not bother, to read emails. They insist that it is best to go and speak directly to someone about any problems.

2. Those who insist on limiting face-to-face interaction as much as possible and prefer documented dialogue (believing it provides better control). They prefer everything that requires decision-making to be sent to them by email. A number of these people I worked with required a priority number/designation to be assigned in the email subject title or referencing system. The priority ranges/notations varied from one Project Manager to the next (e.g. one required a number range [from 1 to 5], another a 'High down to Low' categorisation, and yet another wanted to see emails classified on an 'Urgent or Non-Urgent' basis).

3. Those who ask that all important matters should be dealt with face-to-face on an urgent basis, with an email sent in advance to clarify points if the subject matter is difficult. Short emails were normally the preferred way I saw for handling day-to-day material or a request for specific advice on resolving low-priority issues.

I have found that it is best to establish from the outset which communications route the PM prefers, in order not to cause upset because you could not read his/her mind about how he/she prefers to operate. Sometimes it all boils down to different cultural approaches, so you cannot assume the answer unless you truly know the environment you are operating in. I am not making any comment about those different management styles, other than to say that, for a variety of reasons, I prefer to operate along the lines of alternative 3 above; but each to his/her own is my general motto. Getting the PM's communication route preference fixed at the outset will make for a smoother ride all round. I myself have fallen victim to not getting this right.

6.5 The Project Controls Manager

6.5.1 Range of Functions May Vary

The role/function of the PCM is an area where no two companies seem to espouse the same philosophy. One company may expect the PCM to look after only the Planning/Scheduling activities and prepare the various progress reports. Another company may also want the PCM to be the key person to support the Project Manager at the monthly progress meetings (both internally, within the Contractor's own organisation, and externally, particularly with the Employer). Many PCMs also have the overall responsibility for budget control and dealing with all Variations/Claims (where the Project Contract Manager and his support team are required to report directly to the PCM). I have even known some PCMs be assigned the responsibility of preparing the requisite PowerPoint[15] presentations for Employer meetings and for presenting/discussing them with the Employer. And some companies require their PCMs to undertake a varied mix of those different functions.

There is a great deal of sense in giving one individual overall responsibility for monitoring and reporting on the progress and costing aspects of a Project. However, that approach has to be set up in such a way that the PCM is not swamped with too much 'hands-on' work that he/she must personally do. This means that the support teams need to be adequately staffed with competent people leading them (i.e. Planning Engineers, Schedulers, Budget Controller and support personnel [i.e. Cost Engineers/Quantity Surveyors], Contracts administration and Subcontracts management teams, and Report Writers, etc.). In this way, the PCM can have time to identify what is not going according to plan and develop the necessary corrective action. On the other hand, if the PCM has insufficient time available to vet everything properly, the probable end result will be that the reporting from the other Managers will become sloppy, and control of both the Project Schedule and the Project Budget will become more and more inadequate.

6.5.2 PCM's Primary Functions in the Wider Role

On the assumption that the Board of Directors (or perhaps the Contractor's Corporate Management Team) wishes the PCM to take on the more comprehensive, all-inclusive functions set out above, the following sets out the primary responsibilities of the PCM.

1. Manage the Project Controls Department, and organise, direct and monitor the activities of its personnel.

2. Develop the Project Schedule based on the Work Breakdown Structure (WBS) agreed in conjunction with the Project Manager, the Engineering Manager, the Procurement Manager, the Construction Manager and the Commissioning Manager, and organise the planning/scheduling control activities.

3. Organise the compilation of realistic manpower and construction equipment histograms for sharing with the Project Manager and the Project Director, in order for them to be able to plan for adequate resources to be available on time.

4. Develop the 'Cost Breakdown Structure' in conjunction with the Finance Manager, and organise the budget control activities.

5. Organise the preparation of reports on the Project's progress and cost control status, using appropriate software (such as Primavera for planning/scheduling) in conjunction with other computerised database control systems.

6. Liaise with the following Managers in order to keep on top of the Variations and Claims situation:
   (i) the Project Contract Manager, in respect of Variations and Claims to and from the Employer;
   (ii) the Contract Administration Manager and, if the Project's size and/or complexity is large and warrants an extra pair of specialist hands, the Subcontracts Manager, in respect of Variations and Claims to and from the Subcontractors; and
   (iii) the Procurement Manager, in respect of the impact of late delivery of key materials, goods and equipment on the Project Schedule, as well as Variations and Claims to and from the Vendors.

7. Alert the Project Manager and other Project Management Team members when the Project's progress appears to be behind (or is likely to fall behind) the Project

8. Vet the Project's progress payment applications prepared by the Cost Management Team for submission to the Employer.

9. Vet the payment recommendations prepared by the Cost Management Team in respect of progress payment applications submitted by Subcontractors.

10. Vet the payment recommendations prepared by the Procurement Department in respect of progress payment applications submitted by Vendors.

11. Ensure that Earned Value Management reporting is properly conducted by the Finance Manager's team, and that the conclusions arising therefrom are adequately conveyed to the Project Manager.

6.5.3 Importance of the PCM's Reporting

The content of weekly and monthly progress reports should go well beyond progress/schedule and budget/cost information. It should extend to providing full details of such matters as the KPIs for safety and quality control, etc. Those reports should also include the status of Procurement for all key items and all other important activities, such as factory testing of plant and equipment, etc. In addition, information about the next period's work intentions needs to be conveyed to the Employer's Team. All of that information required for reporting will take time to prepare, collate and be prepared for presentation in an acceptable format for submitting to the Employer. Before such presentation to the Employer, all the contents of the report must be inspected and digested by both the Project Manager and the Project Director for discussion with the PCM, to ensure that all problem areas are fully understood and reasonably under control.

For the foregoing reasons, the PCM must be given full authority over the Managers of all the other Departments when it comes to insisting on being provided with the necessary data for reporting, and the PCM's Report Writers too must be accorded respect. However, the problem often is that the other Managers are too busy 'doing the work' to take the reporting seriously, which can then lead to inadequacies in the formal reporting and a very disgruntled Employer. On behalf of the PCM, the Project Manager must therefore be very strong on this issue from the outset and convey the importance of good reporting to all the Managers, making them responsible for ensuring that the quality of all such reports is beyond reproach.

Failure to handle the reporting properly will inevitably lead to many problem issues (as the Employer's Team sees them) becoming the subject of contractual letters that require a great deal of work from the Contractor to respond to satisfactorily. That will inevitably detract from the work in hand, as well as erode the Employer's confidence in the Contractor. Added to that, poor/inadequate or inaccurate reporting will often be referred to by the Employer's Team when defending Claims received from the Contractor. Typically, the Employer's Team will claim it is a sure sign that the Contractor did not have proper control of the Project's implementation. This is a very big risk that the PCM must therefore take all steps to guard against.

To complete the understanding of the risks associated with the work of the PCM, it is advisable to read the respective roles, functions and responsibilities of the sub-Departments/Teams set out in both Section 6.6 (The Project Controls Department)

6.5.4   Overloading the PCM's Responsibilities

All too often, the PCM is tasked inappropriately with taking responsibility for activities that rightfully should remain the responsibility of others with greater experience. A good example of this is where decision-making about purely contractual issues is pulled under the wing of the PCM, which from my own experience I consider is the wrong approach to adopt. My reasoning is not only that the PCM's technical expertise is very different from that of the specialist Contract Manager, but that such an approach runs the risk that the Project Manager may not be made aware that there are major differences of opinion between those other two Managers as to how problematic issues should be handled.

An additional potential risk is that the PCM will be too thinly stretched if his/her responsibility spectrum is thrown too wide and the corresponding workload increased too much. This caveat also applies to cost/budget management, which is equally a very specialist activity and should remain independent from the Project Controls Department. All too often however, this responsibility is loaded onto that Department to handle (usually ineffectively, from my perspective, based on my actual experiences). However, vetting the outputs from other Departments that are handling payment processing (such as the Procurement Department, the Contract Administration Department and/or the Subcontracts Management Department) is an essential and valid task for the Project Controls Department to perform, since it adds a vital 'checks-and-balances' step into the chain of financial control.

On the other hand, there is nothing wrong in the Project Contract Manager and the Finance Manager reporting to the Project Manager through the PCM, provided that the recommendations of such Managers are faithfully passed onto the Project Manager. If there are differences of opinion in respect of reports, the PCM can always add a rider to the recommendations of those others. The thing to bear in mind here is that there is no point in employing people in key positions if they do not have sufficient experience to do the job properly. Where professionals are engaged, it would also be a total waste of resources (and definitely not good management practice) to allow others with perhaps lesser expertise to ride roughshod over the professional opinions of such other experts.

A balance has to be struck here, especially on large or complex Projects, since the Project Manager needs to ensure that he/she does not operate an 'open-door' policy to every Manager and Sub-Manager, otherwise there may be insufficient time to deal with the more important problems. Making the PCM the 'gatekeeper' through which certain Managers must pass first before getting to the Project Manager can often help save the latter's valuable time. However, there needs to be a safeguard imposed so that the PCM is not allowed to block access to the Project Manager where essential issues are at stake and the opinion of others does not match the PCM's own perception of the matter.

6.6   The Project Controls Department

6.6.1   The Department's Objectives

The objectives of the Project Controls Department (PCD) are seldom understood in the same way by all the other Managers on a Project. This is in no small part due to the fact that the opinions of the various Boards of Directors tend to vary extensively as individual responsible for managing the PCD). It is therefore essential to clarify, as a matter of priority, the objectives of the PCD on each Project, as well as the reporting outputs required from it. This is so that all other Managers will be able to appreciate what activities the PCM needs to have authority over if he/she is to be held accountable/responsible for the reporting outputs.

To be certain that the PCM is being fed with the right information as to the true status of the Project at any given time, the reports from all the various Departments involved need to be checked thoroughly by the members of the PCD. The purpose of such report checking is to see if there are any discrepancies between what one Department is saying compared with what another Department believes is happening. There are many different Departments to take into account, and the relationship of one to another is complex. Where the Contractor is also going to be responsible for the Operations and Maintenance activities, the interfaces between the Departments become even more complex, and the PCD needs to ensure that the cross-checking of a Department's reported facts about the Project's status is coming from the right source. The internal interfaces alone for the foregoing report checking activities can be seen in Figure 6.1, where the oval line depicts the full extent of the PCD's area of responsibility for collecting and analysing the reports of the various Departments and implementation teams.

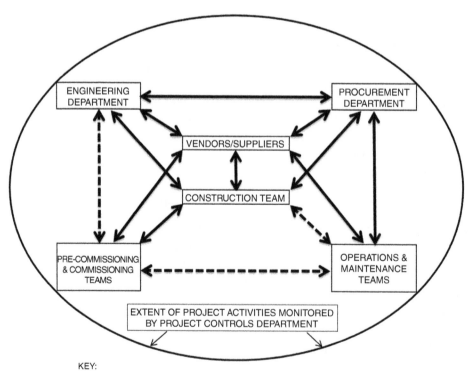

KEY:
SOLID LINES = FIRST-TIER (DIRECT) INTERACTION/DEPENDENCE LINKS
DOTTED LINES = SECOND-TIER (SUPPORT) INPUTS NECESSARY

Figure 6.1 Interfaces for EPC Project's Departments and Teams.

6.6.2   Planning and Controls Explained

Since 'planning' is recognised as being one of the key activities associated with the PCD but 'controls' is the word that sits in the Department's title, before proceeding further it is therefore necessary to be clear about the answers to the following questions.

(1) What 'planning' is done by the PCD for a Project, and why is it necessary?

(2) What 'controls' are exercised by the PCD for a Project?

The following attempts to provide answers to both the above questions.

(1) Planning of all major activities for a Project involves allocating appropriate implementation resources to those activities (manpower and, where applicable, construction equipment), as well as deciding the time that will be needed for each such activity. The planning then needs to be developed further to ensure that, collectively, considering the impacts of one activity upon another, it is possible to reliably check at any given moment that the Project can be completed by the contractually required completion date. On the other hand, if the Project looks set to run into delay, such situation needs to be identifiable early enough for remedial action to be effective for bringing the Project back on schedule. The principal output of such planning work that enables the forgoing to be achieved is, typically, a networked Project Schedule, in bar-chart format, with the Critical Path established and delineated.

(2) The planning exercise would be a waste of time if it could not be possible to see reasonably easily that all key Project activities were proceeding not just according to plan but also in line with quality requirements. To achieve that objective, the PCD needs to establish a worthwhile system of controls to reliably check how each individual major activity is progressing. This is so that, if any one of those activities is seen to be veering off course, such a situation can be spotted in good time, and the erring activity steered back on course again in an effective manner. The primary controls used are sets of appropriate KPIs relevant to each particular activity. These cover such things as the number of drawings submitted for review and approved for construction, the number of Purchase Orders issued, and the statistics for such things as HSE issues, Punch List items, etc. The KPIs need to cover the activities of all Departments, so the totality of the KPIs can be quite extensive, according to how deeply the Project Manager wishes to exercise monitoring and control of work activities at a detailed level.

It should be noted that none of the above activities denotes any controls being exercised by the PCD or the PCM. On the contrary, all the outputs from the PCD should primarily be aimed for consumption by the Project Manager so that he/she will be suitably informed at all times. This should then enable the Project Manager to make the right decisions as to where any necessary remedial action needs to be taken to bring the Project or any specific activities back under control.

The principle issue here is that the PCM has no authority to directly control any of the other Managers, nor is it necessary or advisable for him/her to be given such authority. That authority should rest solely with the Project Manager, failing which it is likely that confusion will reign, and that the chain of command will be misunderstood by other Managers, possibly to the detriment of the Project.

6.6.3   PCD's Planning and Controls in Practice

To achieve the purposes of the foregoing system of controls successfully, the PCD must undertake the following:

1. **Activity Identification**, which involves breaking the Project's activities down in an orderly and structured way, commencing at the topmost level with a short list. That list should encapsulate, in a meaningful and logical way, all the work that needs to be accomplished in respect of the Project's implementation. The list must then be broken down into greater detail at subsequent lower levels, where each upper level is a 'roll-up' of the detail in the lower level. The entirety of the items so listed is referred to as the Project's 'Work Breakdown Structure' (WBS). Great care must be taken to ensure that the final list is not so large as to render it unwieldy when trying to spot progress delays and quickly re-plan later to overcome such delays.

2. **Activity Planning**, which involves (i) carefully considering each item comprised within the lowest applicable level of the WBS, and (ii) identifying the most appropriate resources (manpower and construction equipment) required for the accomplishment of each individual task.

3. **Task Scheduling**, which involves determining:

    (i) the order/precedence in which all the separate tasks need to be undertaken, bearing in mind which tasks need to be completed before other subsequent tasks can commence, and

    (ii) allocating the resources needed for each task and the time required to complete each task.

    Task Scheduling is an iterative and time-consuming process that requires all other Department Managers to get fully involved, in order that the end product (the Project Schedule) will be accepted by the entire Project Management Team as being realistic and achievable. The essential outputs required from Task Scheduling are:

    (a) a written strategy to define the basis of the logic to be adopted in the Project Schedule, including information about working hours, public holiday periods (such as Christmas, New Year and Easter holidays, Eid and other religious-type holidays, etc.), restricted working periods (such as for the Ramadan fasting month, mid-summer or mid-winter working), required overtime working, review cycle allowances, predecessor activities, successor activities, etc.;

    (b) an overall 'networked' bar chart that lists all the activities comprised within the WBS, and which shows the period/duration allocated for undertaking each and every task; and

    (c) a planned progress curve showing the cumulative progress anticipated from day one of the Project through to the planned completion date.

The above description of the Task Scheduling activity has deliberately been pruned down to its bare minimum. The reality is that the activity is extremely complex and time-consuming. It is also fraught with difficulties, requiring the Planner/Schedulers to make decisions for all non-key activities as to whether or not to increase the resources or, alternatively, to eat up some of the Float in respect of those activities. On major Projects, this requires carrying our resource levelling analysis using sophisticated software; it would be impossible to do this manually. Those interested in trying to understand a bit more about the topic (without getting swamped with too much detail) might well appreciate reading Wilkens's conference paper, entitled 'Fundamentals of scheduling & resource leveling'.[16]

4. **Task Monitoring**, which involves plotting the actual progress achieved for each task on a weekly/monthly basis, based on the information made available by all the other Department Managers.

5. **Progress Analysis**, which involves (i) comparing the actual progress achieved for each task against the originally anticipated progress, and (ii) discussing each glaring discrepancy with the relevant Department Manager. This analysis is applicable to all tasks where physical progress is required to be achieved, such as the production of engineering and procurement documents/reports, construction work, commissioning activities, etc.

6. **Performance Analysis**, which involves (i) reviewing the KPIs or other performance expectations against actual performance, and (ii) discussing each glaring discrepancy with the relevant Department Manager. This analysis is applicable to all support activities where data and/or statistics are required, such as Health, Safety and Environmental matters, Quality Control issues, Cost Control activities, Variations/Claims issues, etc.

7. **Weekly/Monthly Status Reporting**, which involves compiling management reports for internal or external use, such as will highlight to the Project Manager all the areas of concern for the reporting period. Sufficient backup data/information should be provided for the Project Manager to be able to make a decision as to where his/her input is required, in order to ensure that the Project will remain under proper and effective control.

6.7 The Planning/Scheduling Team

6.7.1 Unrealistically Short Project Schedules

Although most EPC Project bid submissions will require a preliminary Project Schedule to be attached, there is generally insufficient time allowed for in the bid preparation period for the Contractor's Bid Proposal Team to have thoroughly determined what all the Project's detailed components are. The primary reason for that is that the Detailed

---

16 Wilkens, T.T. (2006). Fundamentals of scheduling & resource levelling. Paper presented at PMI Global Congress 2006 - North America, Seattle, WA. Newtown Square, PA: Project Management Institute. https://www.pmi.org/learning/library/scheduling-resource-leveling-project-progression-8006 (accessed 27 April 2019).

Design work will not even have been commenced at the time the bids are submitted. At best, some preliminary designs may have been supplied by the Contractor at the bidding stage, simply to demonstrate that the bidder was serious about securing the Project award. Logically linking the sequencing of all the anticipated activities to ensure that the Employer's required completion date can be met is therefore generally only indicative at the bidding stage; it is far from a fully developed plan.

When working on the PMC side, I often saw that the Employer had been expecting that the bidders would offer shorter completion periods with a correspondingly lower bid price. However, the bid preparation time had generally been far too short for the bidders to properly prepare detailed alternative bids. Added to that, the time-frame for completion specified in the bids was usually too short, yet most bidding instructions prohibited a longer bid period being proposed. I have therefore witnessed many construction Contracts being signed up with ridiculously short completion periods incorporated. Once the Contract had been signed, the Contractor was thereafter always required to complete the work by the date stipulated therein, no matter how unrealistic the completion time had been. Sadly, that situation is no different today than it was 50 years ago, when I first started out in the construction world.

6.7.2 Importance of a Sound Project Schedule

Notwithstanding that 'Engineering, Procurement and Construction' activities are the main focus of this book, the most successful EPC Projects will have relied very heavily on implementing a good planning strategy to ensure that the Employer's number one objective is met (i.e. completion on time). Although the majority of Project plans start to become obsolete almost immediately after they are tabled, working through the planning process is essential so that the impact of subsequent unexpected changes can be assessed quickly and properly. Where a Project goes badly wrong, it will almost certainly be because the planning strategy was flawed or, most likely, did not exist at all. Compiling a sound Project Schedule is therefore of the utmost importance in order to stand any chance of the Contractor handing over the Project by the contractually agreed completion date.

One major problem I have observed, which arises time and again, is that some Project Managers (and even some Project Directors too) are reluctant to disclose full details about how the Project Schedule was derived. They all seemed to think that giving away too much information would prejudice them when it came to claiming an extension of time. However, I have seen some PMCs refuse to accept the Project Schedule as a legitimate document for measuring progress, simply because insufficient information had been provided to allow them to check the validity of the Contractor's programme. Based on what I saw follow, that was not at all to the benefit of the Contractor. My advice to Contractors, no matter which side of the fence I find myself on, is always to be as open and transparent as possible about how the programme has been put together. This is because, first of all, it is essential in my opinion that the Employer's Team embraces the Project Schedule as a worthwhile document. I believe that aids in locking in the Employer morally to helping the Contractor maintain the Project's time-line. Giving the Employer an excuse to run away from not just accepting but approving the Project

Schedule is a big mistake as far as I am concerned. I am speaking as somebody who has had to deal with very many extension of time and disruption claims on behalf of both Employers and Contractors.

Having said that, I have to add that I have come across an alternative school of thought about the Contractor being transparent in regard to sharing the programming details. That is that the Employer's Team will use the Contractor's declared resourcing and expected productivity levels against the Contractor wherever possible to defeat an extension of time or disruption claim. Certainly, that risk does exist. However, my point is that the Contractor's claim that the Employer caused delay or disruption is so much harder for the Employer's Team to accept if the originally expected production outputs are not made available. Added to that, the Employer's Team may well suspect that the Contractor is withholding the programming data so that it will be easier to play games later if the Contractor desperately needs to secure an extension of time. This is where the Contractor needs to take a commercial decision as to how much programming data to release, which might depend a great deal on the quality of the relationship the Contractor has with both the Employer's Team and the PMC.

6.7.3 The Project Schedule's Owner

When preparing the Project Schedule, the Planning/Scheduling Team must fully understand exactly who the 'customer' is that it is being prepared for. A lot of Managers seem to think that the customer (and owner) for the Project Schedule must be the Employer, but this is most definitely not the case. Indeed, a real problem comes into play if the Employer considers that the Project Schedule is prepared for the Employer alone and must therefore belong to the Employer. This is because the Employer often then tries to start dictating how the Contractor must arrange the Project Schedule, right from the items to be considered (i.e. the activities contained in the WBS), down to the time to be taken for each activity. I have even seen some Employers who thought that re-sequencing the Contractor's access to critical work areas was the Employer's prerogative. Their mistake was that the Project Schedule is not produced first and foremost for the Employer's benefit.

The number one customer of the Planning/Scheduling Team is, instead, the Contractor's Project Manager, because the detailed workings within the Project Schedule belong to the Contractor. The Employer only owns the rights to the contractually required completion date. That is because, on most EPC Projects, that date is usually agreed as being one of the fundamental ingredients of the contractual arrangement between the Contractor and the Employer, along perhaps with a few intermediate milestone dates.

An area of contention that often arises between the Employer and the Contractor is in regard to the ownership of Float. If the Invitation to Bid documents make the statement that all Float in the Project Schedule belongs first and foremost to the Employer, the Contractor must ensure that an objection is raised to that in the bid submission documentation. Failure to do so could result in the Employer arguing that the duration shown in the Project Schedule for each activity cannot be extended without the Employer's approval. That would impose an enormous strain on the Contractor in attempting to keep tabs on the duration of each separate activity to ensure that the original duration for each activity is maintained. In effect, it would make the Project Schedule a contractual document, which it is not supposed to be. Added to that, if the Employer used up all the Float it could mean that completing the Project on time would require a larger slice of good luck than is always needed anyway. This issue is not anywhere near as clear-cut as one may think, and there are many different schools of thought about it, as Bailey observes.[17]

I have found that there is a great deal of confusion in understanding precisely what Float is, so I feel that I should make that issue clear here. I will start with what Float is not: it is not the additional time allowance that the Contractor had built into the duration of an activity to ensure that the it would be completed in good time so as not to delay the start of any other successor activities. The Float for an activity is the amount of time shown in the Project Schedule before it needs to start *plus* the time shown in the Project Schedule before it needs to be finished (together, the Total Float). As Usmani succinctly puts it: 'Total float is the amount of time an activity can be delayed without delaying the project completion date. On a critical path, the total float is zero. Total float is often known as the slack'.[18]

The above means that Float is not per se critical for the Contractor; in other words, the Employer can use up some of the Float without directly affecting items on the Critical Path. It is therefore generally considered nowadays that Float belongs to the Project, and the first entity to require using any Float is the party entitled to use it.[19] That is what the Contractor should stick out for. However, use of Float may become critical for an activity if events occur that eat up its Float to the point where there is no Float at all left for that activity. If that happens to too many activities and the slack everywhere is eroded too far, the Contractor could run into the problem that activities that had plenty of Float but took a little longer to complete have now ended up on the Critical Path and delayed the Project's completion. The Contractor must therefore keep a watchful eye out on exactly what Float is used up by the Employer (on whatever activities it occurs), since that data may prove vital if it is needed to defend against paying Liquidated Damages later down the line.

6.7.4 Managers Must Embrace the Project Schedule

At the outset of the Project, the Planning/Scheduling Team must produce a Baseline Project Schedule that everybody considers is realistic and is committed to achieving. In theory at least, that should make the Project Manager's job much easier. Regrettably, I saw that too many Contractors paid lip service only to implementing a good planning strategy. Instead, a large number of them seemed to place heavy reliance on what I call 'Plan B'. Quite simply, that was to flood the Project with workers later down the line, hoping to make up for the delays caused by planning deficiencies/inadequacies. That approach sometimes became necessary due to the Planning/Scheduling Team having

---

17 Bailey, J. (2014). *The Society of Construction Law Delay and Disruption Protocol: A Retrospective Analysis*. Copyright Julian Bailey and the Society of Construction Law, 14–15.
18 Usmani, F. (updated 9 February 2019). *Total Float Versus Free Float*. PM Study Circle. https://pmstudycircle.com/2013/03/total-float-versus-free-float (accessed 12 April 2019).
19 Society of Construction Law. (2017). *SCL Delay and Disruption Protocol*, 2e, 29, clause 8.5.

issued the Baseline Project Schedule far too late. However, I also witnessed that many Project Schedules were inapplicable and quite useless for the reason that they had not been fully agreed by all the key Department Managers before they were issued. Sometimes that led to the need to re-baseline the Project Schedule month after month, the end result of which was that many Managers lost faith in the Planning/Scheduling Team's ability to drive the Project to completion on time.

Creating a worthwhile Baseline Project Schedule will not be possible if left to the Senior Planner to prepare it with input coming only from the Project Manager. However, I have observed that this situation seems to suit a lot of Managers, because it provides each of them with an excuse if their particular Department later fails to meet the milestone dates linked to their Department's Deliverables. They simply claim that they were not given the opportunity to have a hand in developing the Project Schedule. That 'get-out-of jail-free' loophole needs to be firmly closed. This must be done by the Project Controls Manager (working in conjunction with the Planning Manager), ensuring that all other concerned Managers are involved in establishing and agreeing the Baseline Project Schedule. As a minimum, those Managers must include the Engineering Manager, the Procurement Manager, the Construction Manager, the Subcontracts Manager, the Commissioning Manager and the Handover Manager. It may also help to remind those Managers at the outset that their involvement is needed so that they can then be considered to have assumed ownership of the finalised Project Schedule insofar as it relates to their respective Department's work.

6.7.5 Compiling the Work Breakdown Structure

The first task of the Planning Manager must be to ensure that the final version of the WBS is as good as it can be. It does not have to be a long list of items to be good, and the emphasis should be on the usefulness of the items identified. It must be remembered that, if the Project Schedule contains a huge number of items, many people will not bother to read it properly. They could therefore miss the inter-relationships of items important to their Department's activities.

An excellent way to start off the WBS listing is therefore to try to limit the Level 1 list to no more 20 items, so that they can be fitted onto a single A3-sized landscaped sheet when printed out. That short list can then be easily discussed in depth with all key players present. After the Level 1 WBS list has been firmly fixed, the Level 2 list can be similarly compiled and discussed, followed thereafter by the Level 3 list. Taking these steps will ensure that the WBS will be adequate for the purpose of planning and monitoring the progress of the Project generally, and the Project's detailed activities in particular.

6.7.6 Determining the Planning Strategy

Once the complete WBS has been determined with finality for Levels 1–3, consideration must then be given to the planning strategy to be adopted for allocating the sequencing of all the identified activities (along with their likely durations and start/finish dates). Determining the planning strategy, and then faithfully following it when compiling the Project Schedule, will help to identify where problems in the strategy lie. This step is vital to ensure that the completed Project Schedule will be as useful as possible to all members of the Contractor's Project Management Team. It too must therefore be done in conjunction with the other Department Managers. When that step has been completed, the Department Managers will then be much more likely to refer to and rely upon the Project Schedule when planning and monitoring the activities for their own respective Departments.

An important task that must be done after the initial planning strategy has been determined is to put it to the test. That can only be done after all the details have been input for each of the activities sitting at Level 4, since the resources available for each individual activity will determine the duration of each. Once the initial time-frames for each activity have been established, a Critical Path determination can be conducted. Following that, the Critical Path can then be analysed to determine the amount of Free Float and Total Float available. That is done by conducting both a forward pass and a backward pass through the Critical Path. The correct management of Free Float is essential, as Wilkens rightly observed: '…as long as you manage the free float and don't exceed it, your project is guaranteed to finish on time!'[20]

6.7.7 Finalising the Baseline Project Schedule

Determining the finalised Baseline Project Schedule is an iterative process, and the planning strategy may need to be modified to accommodate any necessary changes found for the sequencing of the activities. This is often found to be necessary after resource levelling has been conducted and resourcing problems thereby discovered. The Planning Manager should therefore organise meetings with all concerned Managers to present the intermediate versions of both the planning strategy and the Baseline Project Schedule to them. The objective of such meetings should be to run through those documents in detail to agree any further modifications that need to be incorporated. It is then important to ensure that all Managers also sign off against the final versions of both documents. Only in this way will it be possible to ensure that those other Managers have properly bought into the finalised Baseline Project Schedule and assumed full ownership of it for their respective areas of responsibility.

6.7.8 Schedules Beyond the Project Schedule

Many of the other Managers (outside of the Project Controls Department) tend to do their own thing with scheduling at the lower, more detailed levels. Very often, they prefer to use spreadsheets for Level 5 and below that are not linked in any way to the Baseline Schedule. This often proves disastrous for the integrity of the Baseline Schedule, since it becomes impossible to monitor properly what the true progress of the Project is. However, one of the principal reasons for this situation occurring is that the Planning Team is often understaffed and is therefore unable to offer the required support to the Construction and Commissioning Teams.

Another major reason for the use of spreadsheets being employed at the lower levels of the Project Schedule is that there is no connectivity between the Site (where the

---

20 Ibid., Wilkens, T.T. (2006). *Fundamentals of Scheduling & Resource Levelling*, Free Float. Paper presented at PMI Global Congress 2006 – North America. Seattle, WA; Newtown Square, PA: Project Management Institute.

day-to-day physical activities are taking place) and the Contractor's Head Office (where the computer server holding the original of the Project Schedule is generally located). This can result in the need to double-handle emailed updates which, more often than not, then leads to severe delays in the updated versions of the Project Schedule being made available at the Site. The only way around this problem is for the Contractor to bite the bullet at the outset, recognising that there will be a continual need to change the implementation plans (which is just the nature of the construction business). This means that a cloud-based software scheduling system needs to be put in place to enable the updated Project Schedule to be available to all Managers on a real-time basis. However, this will still require adequate numbers of planning personnel to be available to respond quickly to the need for updated plans. Waiting until the delays are out of control before trying to put proper schedule controls into place will not save the day, and it will certainly not prevent Liquidated Damages being levied against the Contractor.

6.7.9 Major Risks for the Planning Manager

The major risks for the Planning Manager are as follows.

1. The WBS proves to be inadequate for control purposes. This can occur when bids had been compiled in haste and the Project Implementation Team simply accepted what had been prepared previously. It can also sometimes occur if the Employer's Team has dictated what the WBS should be at the bidding stage, and everybody thereafter simply accepted it without deeper consideration as to the implications. I have seen problems arise where the Employer wished for sections of work within larger sections to be broken out as discrete sections, but where the workflow was intrinsically linked to work in the larger section. It caused all sorts of logistical linking problems that could only be overcome by ignoring the Employer's wishes. Problems can also arise where the Employer's Team insists that Level 4 items must be shown at Level 3 (making the Level 3 list of items far too long and therefore unwieldy). When such issues arise, the Contractor must therefore be very firm, despite any protestations from the Employer. After all, when all is said and done, the Project Schedule belongs to the Contractor, not the Employer.

2. The Baseline Project Schedule is poorly constructed and does not properly reflect the order or timing of the WBS activities. I have seen far too many work programmes that were either based on wishful thinking or were constructed without proper thought given as to what activities needed to be completed before the following activities could begin.

3. Some Managers do not bother to follow the Project Schedule, or they happily change the work sequencing without informing the Planning Manager about the reasoning behind implementing the changes to the previously agreed sequence of working.

4. The progress reporting in some areas is poorly worded or scant (which is sometimes done deliberately), and delay to the contractually required completion date does not become apparent until too late in the day.

The best mitigation measure to put in place to avoid the above risks from turning from theoretical problems to actual problems is for the Planning Manager (through the Project Controls Manager) to involve all other relevant Managers in preparing a worthwhile Baseline Project Schedule. This should be done right from the point of compiling the WBS through to deciding the best work sequencing arrangements, and then having all Managers collectively approve that Schedule; it must be a team effort.

Beyond taking ownership of the Project Schedule's provisions for their respective areas of work, all Managers must also be encouraged to keep other Managers informed if events occur that mean that the sequencing of some work in other Departments also has to be changed. This can only be done by fostering good communication between all the Department Managers and, if the Project Manager is failing to do that, then the Planning Manager must ask the Project Controls Manager to take steps to rectify the situation. Poor communication within the Project Management Team all too often leads to Project failures. In a study conducted by the Project Management Institute it was determined that 'Ineffective communications is the primary contributor to project failure one third of the time, and had a negative impact on project success more than half the time'.[21]

A mistake I frequently saw being made by Project Managers was to not recognise the importance of updating the Project Schedule on a monthly basis to reflect actual progress. This sometimes resulted in the Project Manager not bothering to replace the Planning Manager's support staff when resignations occurred later during the construction phase. The attitude of the Project Managers seemed to be based on the fact that, because they were experienced and knew what they had to do and how long it would take to do it, it was not necessary to spend a lot of time updating the Project Schedule.

What those Project Managers all forgot was that it is a bit like driving a car and arguing that there is no problem if you allow your child to sit in the back without a seat belt, because you are an excellent driver. The problem is that it is the poor driving of others that also poses a risk to your child's safety; and you cannot later remedy the fact that the seat belt was not utilised. It is very similar with construction progress, where the Project is proceeding exactly as expected when suddenly the Employer's Team or an external party causes a major upset. When that happens, the Contractor then needs to demonstrate adequately that the Project had been progressing exactly as planned, and that it is solely the action (or inaction) of others that caused the delay. Trying to construct proof of excusable delay retrospectively takes a considerable amount of time, and often is far from being as convincing as it needs to be. In my opinion, spending money to keep the Project Schedule regularly updated to reflect actual progress is therefore not an expendable luxury on a major Project but an absolute necessity.

Furthermore, again in my opinion, not engaging competent senior personnel to keep proper track of progress will always be the wrong move. It would be tantamount to burying one's head in the sand about the possibility of delays arising from the Employer's side. Yet Employer-side delays are almost inevitable on any Project. The reality is that the huge costs and agony that would be involved in having to defend against paying Liquidated Damages for delay when inadequate records have been kept is a risk not worth taking.

21 Coreworx. (20 April 2017). *Poor communication leads to project failure one third of the time.* http://www .coreworx.com/pmi-study-reveals-poor-communication-leads-to-Project-failure-one-third-of-the-time (accessed 17 January 2018).

6.7.10   Retrospective Compilation of Project Schedule

Over the years, I witnessed many tens of millions of dollars expended by Contractors on engaging outside consultants to retrospectively construct a Baseline Project Schedule. That needed to be done because a worthwhile Project Schedule had not been in existence prior to that. Each time this belated activity was undertaken, the primary purpose had been to provide the Contractor with a tool that would enable a forensic analysis of the delay situation to be carried out. All of that was in an effort to compile a worthwhile extension of time Claim. The major problems with having to do that will always be the same:

1. It will take a very long time to do all the necessary research work, compile the Schedule and then conduct the detailed analytical work to prove that the Employer is responsible for the delays encountered.

2. Because of the length of time taken to prepare and submit the Claim, the Employer's Team may well reject it without reviewing it, on the grounds that its submission is 'out of time'.

3. The Employer's Team may play for time and demand that the Contractor submit not just full supporting details before the Employer's Team will even begin to review the Claim but also to supply increasingly more detailed levels of substantiation. This can seem like a never-ending undertaking for the Contractor and, consequently, it can become extremely wearing for its Management Team.

4. The philosophy forming the basis of the Project Schedule may be:

    (i) ripped to pieces by the Employer's Team as being unrealistic, or

    (ii) denigrated on the grounds that it has been constituted specifically to take advantage of the Employer's delays, while at the same time masking the effect of the Contractor's own delays.

In all such cases I observed that, if the Contractor had taken steps to prepare a Baseline Project Schedule right at the outset, then the exercise of proving entitlement to an extension of time would have cost far less in the long run. It would also have cancelled out most of the negative arguments of the Employer's Team and thereby taken far less time to get resolved (whichever way that finally turned out to be). Compiling an adequate Baseline Project Schedule right from the Project's outset (and keeping it properly updated) is therefore an unquestionable must in my book.

6.8   The Engineering Manager

6.8.1   Primary Responsibilities

The following sets out the primary responsibilities of the Engineering Manager (EM):

1. Manage the whole of the Engineering work for the Project, including those portions/elements that are intended to be done under specialist Subcontracts or Procurement packages (for example, towers for overhead transmission lines or catalytic converters for use in refineries).

2. Where the Contractor does not have its own in-house Engineering Team, monitor and chase up the work of the external (third party) team undertaking the engineering design work on behalf of the Contractor.

3. For all Engineering work, coordinate with the Employer's Team (which may well be represented by a PMC) as well as the Contractor's Procurement Manager, Construction Manager, Commissioning Manager and Operations Manager, so that the requirements from all the various parties are fully understood and fully satisfied.

4. Manage all Engineering staff to deliver design drawings and technical documents in full conformance with the contractual requirements.

5. Ascertain the dates by when Materials Requisitions and Technical Bid Evaluation Criteria need to be ready for the Procurement Department to issue Requests for Quotations for critical and Long-Lead Items. This can be achieved by ensuring that those work packages are all allocated to specific Engineers given the responsibility for getting all the documentation together on time (i.e. dedicated Work Package Engineers).

6. Ascertain the dates by when early-start on-Site construction information is required, and ensure that those work packages are also all allocated to specific Work Package Engineers.

7. Ascertain the dates by which information is required for Subcontract bidding purposes, and appoint appropriate personnel to ensure that all such information will be made available by the required dates.

8. Ensure that no aspect of the Engineering work falls behind the time-line indicated in the Project Schedule, and arrange for appropriate catch-up plans to avoid/overcome delays, paying special attention to ensuring that all Vendor Documents/Deliverables are reviewed and processed in good time.

9. Handle design changes arising from the Employer's instructions or unexpected on-Site conditions, but only after discussion with and agreement of the Project Controls Manager and Project Contract Manager. This must be done in accordance with the appropriate written contractual procedures (principally the 'Management of Change Procedure') and notified to the Project Manager.

10. Manage/monitor the activities of the Project's respective Document Control Teams, and ensure that the Employer's required Information Management System is followed and complied with at all times, including the requirements for tagging and any Maturity Management reporting that may be required.

11. Work closely with the Project Information Manager to ensure that all tagging information is being made available and worked on, so that inputs into systems such as IBM's Maximo (or any other Employer like-minded software systems) are completed in a timely manner, and without any fear of delaying commissioning or start-up activities.

12. Ensure that the As-Built Drawings are completed as soon as possible and that they accurately reflect what is shown on the corresponding approved Red-Line Drawings.

6.8.2 The Engineering Department's Customers

The first thing that any EM must be clear about is exactly who the 'customers' are for the services that are being provided by the Engineering Department. A common mistake I observed was for the EM to consider that the Employer of the facility to be built was the Engineering Department's primary direct customer. However, the reality is that the Employer is the *Contractor's* direct customer, not the Engineering Department's direct customer.

The Engineering Department's direct (first-tier) customers are the Procurement Department and the Construction Department. Its second-tier customers are, not in any prioritised order: the Employer's Team, the Vendors, the Pre-Commissioning and Commissioning Teams and the Operation and Maintenance Teams (whether or not the Contractor is responsible for undertaking the operation or maintenance activities). It is true that the Engineering Department must produce its information in full conformance with the Employer's requirements, but if it does not do so then it can never be in a position to satisfy the requirements of the Procurement and Construction Departments. Once the EM has grasped the importance of this point about who the primary customers of the Engineering Department are and guides the members of the Engineering Department accordingly, the more focused the Engineering Department will become on satisfying the requirements of the right body of people first and foremost.

Putting the above point another way, if the Engineering Department does not deliver its outputs to satisfy the real needs of the Procurement and Construction Departments, then that would compromise the chances of the Project being completed on time and could thus result in the Employer's primary objective (i.e. completion of the Project on time) not being met. Not meeting the Employers' primary objective would be the very opposite of prioritising the Employer's satisfaction. However, where the engineering design work was subcontracted to external third parties, I saw that this concept of the primary customer being the Contractor's Procurement and Construction Departments was not so readily recognised by the engineering design Subcontractors.

Once the Engineering Department is properly focused on who the priority customer is at any given stage, the need for greater communication with that customer will be better understood and acted upon. That, in turn, will only be beneficial for the Project because it will ensure, for example, that the Procurement and Construction Departments will be better informed as to *what* Engineering outputs will be coming and *when* they will be coming. That will enable those Departments to plan their respective workloads, and utilise their resources, more efficiently.

6.8.3 Key Preparatory Activities for Engineering Work

In order to be able to meet the needs of the Procurement and Construction Departments satisfactorily, as well as those of the Employer, the EM must ensure that all the following key preparatory activities for the Engineering Department's work are taken care of properly:

1. The total number of documents that must be produced by the Engineering Department over the course of the whole Project, as well as the resources necessary to

produce them, have to be established to a high degree of accuracy if there is to be any chance of tracking the Engineering progress effectively.

2. The documents and associated resources should also be broken down into the following categories (where such are applicable) in order to make it easy to identity quickly if something is not going according to plan:

   (i) Procedures related to Engineering Activities (such as Management of Change, Engineering Quality Assurance, etc.),

   (ii) Conceptual Design inputs (if any),

   (iii) Front-End Engineering Design (where/if this is the Contractor's responsibility),

   (iv) Detailed Design,

   (v) Procurement Support Documentation (such as Technical Bid Evaluation strategies, Materials Requisitions and Technical Bid Evaluation Reports), with Long-Lead Items split out into a separate category,

   (vi) Vendor Document Reviews,

   (vii) As-Built Drawings, and

   (viii) the Engineering Close-Out Report.

3. All such required documents must be compiled into a comprehensive list known as the Master Document Register (MDR). The MDR itself should be split into useful sections, such as the Technical Documents Register (to identify all those documents required to be prepared by the Engineering Department) and, similarly, the Procurement Documents Register.

   The list of the required technical documents will most likely include the following:

   - Updated 'Basis of Engineering Design Document',
   - Process Flow Diagrams,
   - Piping and Instrumentation Diagrams,
   - Plot Plans/Layouts,
   - Block Diagrams,
   - Single Line Diagrams,
   - Engineering Calculations,
   - Drawings,
   - Materials Reports,
   - Specifications,
   - HAZID/HAZOP Reviews,
   - SHEAMS Registers,
   - Materials Requisitions,

- Data Sheets,

- Supplier Documentation Requirements Lists,

- Design Administration Documentation (such as procedures related to the technical design work),

- Materials Take-Offs,

- Spare Parts Lists,

- Technical Bid Evaluation Reports, and

- Engineering Progress Reports.

4. All the foregoing documents should be divided into the various stages of the design work involved, namely the Conceptual Design, Front-End Engineering Design, Detailed Design and Close-Out stages, as applicable. Under those stages, the work should then be further broken down into the various Disciplines involved, according to the Project's various components (such as Process, Pipeline, Piping, Mechanical, Electrical, Instrumentation, Communications and Civil Work).

5. The details of the anticipated time to be taken to prepare all the above outputs must be established too, broken down separately against each line item in the MDR, and further broken down into the different grades/levels of personnel required (such as Lead Electrical Engineer, Senior Electrical Engineer, etc.).

6. Those of the above documents that require review by the Employer's Team also need to be identified, along with the applicable duration of the review periods for each document; those documents that might potentially require re-reviewing need to be identified too.

7. The ready-on-Site dates required for all equipment items (both the Long-Lead Items and the bulk materials), and the out-to-bid dates of the procurement documentation for all those items, also need to be established. That should be done through liaison and in-depth discussion with the Construction Manager and the Planning Manager, preferably with both together at the same time. The outputs of those discussions should then be conveyed promptly to the Procurement Manager, which might then initiate an iterative discussion process between all three Managers to ensure that all potential problems are properly identified and resolved satisfactorily.

8. The Project Schedule must properly incorporate all the Engineering time-frames required to meet all the deadlines established by attending to the above. It must also be broken down such that the impact of a significant change for any item on all other items can be readily assessed (whether it is a time saving or a delay). This also requires the EM to determine that all intervening actions preceding the next steps in the Engineering process are properly integrated (such as the Employer review periods, HAZOP Review Meetings, the bid periods for Vendors and Subcontractors, review of Vendor Documents/Deliverables, etc.).

9. The teams required to form the 'complement' for each design Discipline also need to be established, in order to be able to complete the various activities comprised in the Engineering work within the allotted time-frames.

6.8.4   Primary Risks for Engineering Work Activities

By ensuring that all the above matters are attended to, the EM will be in a good position to identify where the primary risks for the Engineering work sit, and determine what mitigation measures need to be put in place to reduce the negative impacts should those risks materialise. In that regard, the following lists some of the typical risks to Engineering work activities, along with the appropriate mitigation measures for containing those risks.

6.8.4.1   *Late Delivery of Engineering Documentation*

In my view, the greatest sin of all for the EM is for his/her Department to fail to deliver the individual batches of critical Engineering documentation on time. That is because such failures lead to serious knock-on delays for both the Procurement work and the Construction work activities. However, I have observed that this problem is not just a regular occurrence; it is the norm. It does not matter whether we look at internal Engineering Teams or subcontracted Engineering work, the problem always seems to be the same – inaccurate reporting on the actual progress of the Engineering work, leading initially to complacency and ultimately to serious delays for the Project.

To counter this annoying and disruptive tendency towards over-optimism on the part of the Engineering Team, monitoring of the required targets for the Engineering outputs should be done more thoroughly. The best means to do that is to hold daily task meetings as well as weekly wrap-up and look-ahead meetings, in order to see what is happening in detail. In this way, any problems in meeting the required Engineering targeted outputs will be identified early on. That will then allow remedial steps to be taken to catch up any lost progress and avoid repetition of the problem in the future. I agree that avoiding taking those mitigation steps will make the Engineering Manager's initial workload lighter. However, the price for adopting such a laid-back approach will inevitably have to be paid later, possibly when it is too late to redress the delay problem satisfactorily.

There will be many Project Managers and EM's who will object to applying the above mitigation measures, usually on the grounds that they are too costly and time-consuming to implement. That cry of despair will be especially loud from any third-party organisation to whom the Engineering work is sublet. However, as Deloitte observed, many capital Projects in the Oil & Gas Industry considerably over-run both the planned completion date and the budget.[22] There is therefore every reason to ignore the opinion of such people, unless they can demonstrate that all the Projects they were responsible for that did not employ such mitigation measures were nonetheless completed in line with the following requirements (although I do not expect many people being able to trumpet such successes):

   (i) they finished reasonably on time (i.e. with less than a 15% time over-run), and

   (ii) they finished reasonably within budget (i.e. with less than a 10% increase over the planned expenditure).

---

22 Deloitte. (2015). *Oil and Gas Reality Check 2015 (A look at the top issues facing the oil and gas sector)*, p. 18, para. 1.

6.8.4.2 *Gradual Scope Creep*

Gradual scope creep is the term given to the situation where the amount of work to be done increases without anybody becoming aware of it until too late in the day. This happens primarily because it is no more than a slow, almost imperceptible trickle of additional items. Scope creep can add to the cost not only of the physical on-Site work but also quite significantly to the amount of work that the Engineering Department must undertake. Such extras can arise in the following ways:

(1) Small items of work are added in by the Employer's Team, often under the guise of being required in order to satisfy safety concerns.

(2) No apparent new items are added, but positional changes of buildings and equipment are required by the Employer's Team that have the effect of adding small quantities of new work in many areas.

(3) Additional unnecessary initial detailing is demanded, especially where the Employer's Team is required to make a choice between alternative design proposals (e.g. piling versus caissons or soil replacement).

Each Lead Discipline Engineer must be made responsible for reporting to the EM all significant design changes as soon as they become aware of them. To ensure that this is done in an organised fashion, it is best that weekly engineering progress meetings are held between the EM and all Lead Discipline Engineers, where all such extra items and any changes found necessary to the MDR can then be discussed in full. Where additional work is considered to be the result of scope changes instructed by the Employer's Team, the EM must take immediate steps to raise a change order request with the Employer.

6.8.4.3 *Inadequate Monitoring of Engineering Subcontractors*

Where the Contractor outsources the Engineering work to an Engineering Subcontractor, the same principles mentioned above should be applied, except that an added mitigation measure should be for the Contractor to employ its own Lead Discipline Engineers to monitor and report on the work of the Engineering Subcontractor. Some may see this as an unwarranted additional expense. However, the experience of the majority of the EPC Contractors I have been involved with is that most Engineering Subcontractors do not take adequate steps to ensure that the designs produced will keep construction costs to a minimum. Often, additional items are added in simply because it is the easiest way to keep the Employer's Team happy. One way to reduce the chances of this happening is to insist that the Employer's personnel and the Engineering Subcontractor's personnel must not be in direct communication with each other and must communicate only through the EM.

Further, the progress reported by Engineering Subcontractors during the course of the work is all too often overstated. This then frequently shows up only much later as large delays on the Project, but which could have been avoided if a weekly sanity check on the Engineering work progress had been taking place, conducted by an independent party. Essentially, the Contractor's own team of Lead Discipline Engineers can be considered as representing such an independent party. Replacing competent Lead Discipline Engineers with cheaper, inexperienced Engineers will cost the Contractor more in the long term than the reduced salaries could ever save, and would therefore be nothing more than false economy.

6.8.4.4 Disruption of Information Management Tasks

The amount of information generated for even a small EPC Project (comprising both documents and data) can be substantial. In days gone by, before the advent of personal computers, a small army of people worked away in the background to handle all that information, diligently carrying copies of documents and drawings from person to person, and filing the originals for easy retrieval if and when required later. Those cheap support teams of the past have long since gone. Nowadays, almost all information for use in the construction industry is produced electronically, and stored in central locations (on computer servers) for access by relevant team members. Such access can be made available for viewing not only when a drawing or document has been finished but even when it is still a 'work-in-progress'.

Establishing the best means of storing and disseminating information is vital to the success of every Project because, if the flow of information becomes disrupted, it will inevitably result in time being lost. That could then lead to financial losses being incurred by the Contractor. A one-size-fits-all approach to Information Management will not work. If an enterprise-wide Electronic Document Management System (EDMS) is not employed in the company, nowadays it will be essential to appoint a competent Project Information Manager to help set up the most appropriate Information Management System for the Project. That same person should then be held responsible for ensuring that the system runs smoothly and efficiently.

6.8.4.5 Failure to Rationalise the Information Management System

There will be many participants working on a construction Project who will require rapid access to completed information prepared by the Engineering Team; not just to the drawings and specifications but to the equipment data too. The main Engineering Team itself may be an external Subcontractor in some cases, which can make such access difficult for the Contractor's Team unless steps are taken to deal with that problem. The other Project participants interested in such information extend well beyond the Contractor's Engineering Team to such groups as the Project Controls Team, the Procurement Team, the Cost Management Team (particularly the Cost Engineers and Quantity Surveyors), the Construction and Commissioning Teams, the Vendors' Engineering Teams, and the Employer's Team members. For this reason, it is imperative to ensure that an adequate software platform is available for all those Project participants to be able to access the documents and data in real time.

I have worked on Projects where the Contractor's Engineering Subcontractors had their own individual (privately accessed) EDMS, and so too did the Contractor and the Employer's Team. Each system was different from the other, and each contained elements/modules that were not compatible with the systems the other parties were using. Engineering Information therefore had to be double-handled from one system to the other, sometimes in a manually reworked format, which resulted in delays as well as confusion (especially when it came to handling Vendor documentation) and, sometimes, errors occurring.

My conclusion is that, in order to make life easier and more cost effective for the Contractor, it is best if the Contractor invests in a world-class EDMS to handle all Engineering data that all external parties can then access without difficulty. The responsibility for inputting to the Contractor's EDMS should then be placed on those external parties and, if the Employer's Team requires downloading of the data into its own separate system for its own use in the operational phase, then it should be up to the Employer to arrange for that at the Employer's own expense. This must be dealt with at the bid negotiation stage, and the Contractor's bid should be qualified to state that uploading to any separate EDMS sitting on the Employer's server has not been allowed for. Hoping that this major issue can be sorted out in the Contractor's favour after Contract signing would be a disaster for the Contractor. This is because most Employer's Teams I have encountered only wish to make life easy for themselves. They have no interest in what the Contractor's information-handling problems are, seemingly believing (erroneously) that the Contract Price of EPC Projects must cover all risks of any nature. The risk of major delays occurring owing to inadequate information-handling arrangements is huge on a major EPC Project and therefore needs to be eradicated right from the outset.

6.8.4.6   *Reliance on Document-focused Progress Reporting*

One of the ways in which all Engineering Teams fool themselves about the progress made is to report progress based on the number of documents issued, even though many of those documents are incomplete and have a substantial number of hold points on them (usually representing equipment items where Vendor data input is required). The way around this problem is to conduct progress reporting based on a structured Maturity Management System. That first requires each item of equipment to be identified and given its unique tag number and, secondly, to allocate 'gateways' that will aid in determining the design progress of each item. Each individual design gateway is allocated a percentage weighting that is earned only when the particular gateway has been passed. The stage of design reached for any particular item of equipment is dependent on the specific gateways that have been passed through. The typical gateways for an item (or a group of identical items) are things such as (but not necessarily limited to) the following:

   (i) the item has been registered, along with all its characteristics,

  (ii) the Materials Requisition has been fully prepared,

 (iii) the Request for Quotations has been issued,

 (iv) the Technical and Commercial Bid Proposals have been received,

  (v) the Technical Bid Evaluations have been completed,

 (vi) the Commercial Bid Evaluations have been completed,

 (vii) bid negotiations have been concluded,

(viii) the Purchase Order has been placed,

 (ix) manufacturing has commenced,

  (x) manufacturing has been completed,

 (xi) the item has been shipped,

 (xii) the item has arrived at the destination port,

(xiii) the item has cleared Customs,

(xiv) the item has been Delivered to Site

(xv) the item has been installed and mechanically completed,

(xvi) the Operating and Maintenance Manuals have been completed, and

(xvii) the item has been commissioned.

It goes without saying that all Engineering Teams (and some Procurement and Construction Teams too, but not usually the Commissioning Teams) detest such a system of progress measurement. This is because, being 'object' focused and not 'document' focused, it is extremely effective at highlighting precisely where not only design work but other essential activities are falling behind schedule. It is impossible to demonstrate progress as adequately as the Maturity Management System does using the traditional 'number of documents issued' route to determining progress (especially design progress). This advanced method of progress measurement is also very effective at stopping over-assessment of the evaluation of work actually completed. In my opinion, the argument that it takes up too much time to operate does not hold water, as the majority of the information is either usually already available for other administrative purposes or will be required eventually. Additionally, the information recorded within the Maturity Management System can be utilised for other reporting systems (such as for Information Management purposes).

6.8.4.7 Problems Caused by Employer's Non-readiness

There are two ways in which the Employer's Team can make life difficult for the Contractor, and they are by not being ready with:

(i) the document numbering system requirements, and

(ii) the tag numbering requirements.

Having to re-issue documents later due to Employer revisions in either one of those numbering systems causes no end of confusion. It even opens up the door to the Employer's Team possibly re-reviewing resubmitted documents (that have the revised document or tag numbers on board) as if they were first-time submissions. The Contractor must ensure that the Contract Documents fully cover the point that the Contractor will be entitled to an extension of time, with associated costs payable by the Employer, if the Employer's Team and systems are not ready in time. Such readiness must extend to providing the Contractor with all the essential information and data requirements to enable the Contractor to commence submitting its Deliverables without hindrance as a one-time only exercise.

Another problem area is where the Employer insists that tag numbers can only be used in approved batches (meaning that they must first be verified by the Employer's Team). The Contractor must therefore ensure, before proceeding, that the document numbering and tag numbering systems/requirements are fixed permanently ('written in stone') before commencing with the Engineering work. The Contractor should also insist that the control of tag numbers must rest with the Contractor, not with the Employer's Team. This may require the Contractor having to send weekly advice to the Employer's Team as to what tag numbers have been added in the previous week for checking purposes. But that is far better than allowing the Employer's Team to control the tag numbers. However, there may be a valid reason as to why the Employer wishes to maintain tight control of the tag numbering. If that is the case, then the Contractor should endeavour to ensure that no hold-ups will occur to the Engineering Team's work due to having to wait too long for batches of tag numbers to be forthcoming from the Employer's side.

While on the topic of tag numbers, I have seen that actually placing the tag numbers on the drawings is another area where big problems can arise. This occurs because many Engineers seem to hate doing the job, primarily because it inevitably slows them down in completing the design work. On the other hand, many Employers insist that a drawing without tag numbers on is incomplete and can be rejected. From my experience, leaving the tag numbering until later in the day makes it an even heavier task that nobody then wishes to do. The EM must therefore insist that each Discipline Manager will be held responsible for ensuring that tag numbers are put on the drawings (and also faithfully recorded as having been used) immediately it is possible to allocate correct tag numbers. This requirement must apply to Engineering Subcontractors as well as to Vendors and Specialist Subcontractors. I have noticed that strong support from the Project Information Manager will sometimes be needed to help reinforce this requirement.

6.9 The Procurement Manager

6.9.1 Primary Responsibilities

The Procurement Department will usually be the biggest spender by far on a Project, and will often be responsible for setting up Purchase Orders and Subcontracts totalling upwards of 50% of the net Contract Price (i.e. the Contract Price minus the Contractor's allowances for Overheads and Profit). The Procurement Manager therefore has a huge responsibility to ensure that competitive pricing is evidenced for all procured items/services through a valid bidding process, and that personnel employed within the Procurement Department act with honesty and integrity at all times.

In addition, the Procurement Manager must ensure that:

(i) contracts are only entered into with reputable and trustworthy organisations,

(ii) the quality of all goods and services meets the specification requirements,

(iii) timely delivery of the goods (in first-class order) is achieved, and

(iv) the Vendors' on-Site support services are provided as and when needed.

The following sets out the other primary responsibilities of the Procurement Manager.

1. Manage the Procurement Department and organise, direct and monitor the Procurement Team members to arrange for (i) the purchasing and delivery of all materials, goods and equipment needed for the Project without any unnecessary delay, and (ii) organise the bidding and award process for those elements of the Project that are going to be subcontracted. In regard to this latter activity, I have seen that some Contractors approach this matter differently and set up a separate specialist team to handle this work (the Subcontract Procurement Team).

2. Coordinate with both the Engineering and the Construction Departments to understand and comply with the priority of materials, goods and equipment

needed for the construction work. Thereafter, continuously chase up on all outstanding Materials Requisitions from the Engineering Department to ensure, as far as can be managed from the Procurement Department's side, that they are issued in a timely manner along with all necessary Data Sheets.

3. Oversee the preparation by the Buyers of the Requests for Quotations and associated documentation for all potential Purchase Orders.

4. Ensure that evaluation of the incoming bids/quotations is handled without delay, that the Engineering Department does not delay any Technical Bid Evaluation Reports, and that Purchase Orders and Subcontracts are awarded in a timely manner (and in full accordance with the Contractor's bidding/award rules and procedures).

5. Ensure that all materials, goods and equipment are delivered to the Site as per the Project Schedule, especially Long-Lead Items and critical items, including organising the shipping arrangements, appointing the Customs clearance agents and road hauliers, and arranging the insurances for transportation purposes.

6. Ensure that the Construction Department is notified of the requirement to provide adequate warehouse facilities, outside storage facilities and laydown areas, so that materials, goods and equipment are stored in accordance with the supplier's requirements and warranties are thereby not voided.

6.9.2 The Procurement Department's Prime Customer

The Procurement Manager must firmly fix in the heads of all members of the Procurement Department that their prime customer is the Construction Department. The principle requirements that the Procurement Department must fulfil for the Construction Department are (i) to deliver the correct materials/goods to the Site at the right time, and (ii) to ensure that all Subcontractors are appointed in good time. Along the way there will be many things that could go wrong in the Procurement process that could compromise those objectives. One of the Procurement Manager's most important roles is therefore to identify what the prime risks are, and how to mitigate them in the most effective way. This is because delay in the delivery of materials/goods or the late start of Subcontractors has the potential to cause the Contractor to overshoot the contractually obligated completion date. That could then entail the Contractor having to pay the Employer Liquidated Damages for delayed completion.

6.9.3 Deciding Equipment/Deliverables Delivery Dates

The people who will always have the greatest impact on deciding what materials, goods and equipment need to be delivered (and by which dates) are the Commissioning Manager and the Construction Manager. Their initial decisions have to be made completely independently of what the manufacturing and delivery times are, or how long the engineering work will take; the impacts of such things can be deliberated after the list has been compiled. Such a list must then be circulated to the Procurement Manager, the Engineering Manager and the Project Manager for preliminary consideration. Following that, a team meeting should be held to prioritise the order in which the Engineering Deliverables for Procurement bidding purposes need to be completed, and also the dates by when those Engineering Deliverables should be available. Similarly, those same Managers should be the ones to decide, collectively, when the Subcontractors need both to commence and finish their work.

6.9.4 Considerations for Long-Lead Items

A special list of Procurement items that will be labelled as Long-Lead Items (LLIs) will always arise from the foregoing activities, and yet each of the individuals involved in preparing the list will have in his/her mind a different definition at the outset as to what exactly an LLI is. This is because, in construction Projects, items that are not necessarily 'long-lead' in terms of manufacturing and delivery time may well be required early in order to allow other follow-on trades to start and complete their work on time.

One example of this would be a large-scale pipe rack necessary to carry a stack of pipes over a long distance, where the installation time for the pipes and associated instrumentation is also very long, thus putting the pipe rack on the Critical Path (not only the construction work but also the steelwork deliveries). Another example would be the construction of a prefabricated steel wharf, where large quantities of feedstock for the Project will be needed for testing purposes prior to commissioning of the whole facility, and the only cost-effective method of delivering that feedstock would be by ship. This would put the structural steel fabrication, its delivery to the Site and its installation on the Project's Critical Path. Essentially, therefore, all purchasable items for activities that sit on the Critical Path must be considered to see if they are to be classified either as LLIs or urgent/early purchases (which should be given the same priority as LLIs).

The staff members of the Procurement Department of an EPC Contractor are often under continual stress, due to the requirement to get all the materials, goods and equipment to the Site at the right time for the Construction Team. However, before the Buyers can place a Purchase Order for any given item, all the necessary documentation must be made available to the Procurement Department by the Engineering Team, and all prequalification procedures must have been completed. Those goods known as 'bulk materials' (such as standard electrical cables, conduit, piping, etc.) are not generally the cause of delays. Equipment items, especially those identified as LLIs, are a completely different story.

LLIs are very often not 'off-the-shelf' items. On the contrary, they may require special design and be 'one-off' productions. If the Engineering work is delayed beyond the time-frame allowed for in the Project Schedule for any LLI, then it will almost inevitably result in Project delay, since most LLIs sit on the Critical Path (and so will have no Float). Unless the Employer has seen the problem in advance and taken the step of pre-ordering the major LLIs, then the Contractor will be up against the clock from day one. In 20 years of working on EPC Projects, I have only been on one Project where the Employer was foresighted enough to order the critical materials in advance (many hundreds of kilometres of very large diameter gas linepipe). I have therefore experienced late delivery of more LLIs than I care to remember. In every single case of late delivery, the Procurement Department had done everything possible to make up for lost time, but LLIs seldom offer the luxury of a time saving fix. In short, once an LLI late delivery causes a delay on the Critical Path, there is next to nothing that can be done to make up for the lost time. LLIs must therefore be given topmost priority by the Procurement Team.

6.9.5 Late Engineering Deliverables

One of the major risks for the Procurement Department is the same on every Project, large or small: the inputs needed from the Engineering Department may arrive in the Procurement Department far later than required. There will be many unique reasons as to why the Engineering Department is late with those deliveries. However, the first task that the Procurement Manager must undertake is to establish precisely what Engineering Deliverables are required for procurement purposes, as well as when each package of design information has to be made available by.

It is not the prerogative of the Engineering Manager to decide the dates when each batch of Engineering Deliverables will be sent to the Procurement Department. It is the right (and obligation) of the Commissioning Manager, Construction Manager and Procurement Manager to make such decisions collectively. Nonetheless, I have observed that many Engineering Managers seem to want their Department to do its work independently of the demands of others, and to undertake the production of its documents in the order that best suits the Engineering Department's personnel. Sometimes that is done without proper regard to what is actually required to complete the Project on time. Such thinking must be quashed from the outset by the Project Manager (on behalf of the Procurement Manager). That is because, as I have seen many times, if the Engineering Department is allowed to do its own thing and complete its design work as and when it wishes, then the Project will inevitably become severely delayed.

6.9.6 Risk of Corruption Occurring

Not least because of the large amount of money involved in the purchase of goods and services for a major EPC Project, there is more risk of corruption occurring as a result of Procurement activities than in any other aspect of a Project's work. The Procurement Manager must therefore take steps to ensure that the following points are fully understood and faithfully followed by all members of both the Procurement Team and the Engineering Team:

1. Responses to all bidders' queries must be formally issued to all bidders through the relevant Buyer in the Procurement Department. No Engineering Department personnel must be permitted to contact the bidders directly but, instead, must operate only through the relevant Buyer.

2. Competitive bidding and its attendant confidentiality requirements are of the utmost importance, in order not to destroy the integrity of the bidding process.

3. The rules for not opening sealed bids ahead of the formal opening ceremony (where every activity must be formally recorded) must be strictly adhered to. If this rule is flouted, the Employer's Team may insist that re-bidding is required, thereby losing vital time for the Contractor and costing more money to close the same task.

The above points are very important, because any irregularities in the bidding process may give rise to one or more of the bidders calling 'foul', which might then involve international compliance authorities conducting investigations. That could then have dire negative consequences for the Contractor in adhering to the Project Schedule. In order to maintain the confidentiality of the bidding documents, particularly the Commercial Bid Proposals, all Procurement activities for the Project should ideally be undertaken in a secured area, with access restricted to only those given direct responsibility for handling the bidding process for the various work packages. Even with the Technical Bid Proposals, the Procurement Manager must take all reasonable steps to ensure that only those authorised to handle them for technical evaluation purposes are given access, and that all such documents are securely locked away at the end of each working day. The primary reason for this is that some potential Vendors and Subcontractors may seek to bribe the Contractor's staff to pass on specific details about the contents of another bidder's technical proposal.

One of the ways to maintain closer control of the soft copies of bid submittal documentation is for the Procurement Department to have its own computer server (linked to the Project's main server, of course), which then blocks off any opportunities for copies to be sent outside of the Procurement Department. A further step is to ensure that no electronic copies of data can be made, and that can be achieved by not providing individual standard desktop computers at the individual employee's workstation and, instead, utilising a 'Thin Client' solution.[23] Further, no commercial documentation received from Vendors and Subcontractors should be allowed to go out of the Procurement Department unless signed for by a responsible individual on each occasion. Taking all these steps collectively will greatly reduce the risk of corruption surfacing in the Contractor's Procurement process.

6.9.7 Tracking Manufacturing and Delivery Status

One of the biggest problems for the Procurement Manager is keeping track of precisely where every item stands in relation to its required ready-on-Site date. This is especially true when the list of items to be procured starts to become large (over 50 items). Using a proprietary computerised tracking system can significantly help to overcome this problem. Such a system will be able to produce reports that show the progress status of every Purchase Order required, complete with the following sorts of information:

- bidder prequalification requirements,
- bidder prequalification status,
- ready-on-Site requirement,
- shipping date requirement,
- Materials Requisition availability,

---

23 Thin Client – a computer that lacks its own hard drive and must therefore run on application software resources stored on a remote server.

- Request for Quotation issuance status/date,

- bid return date,

- Technical/Commercial Bid Evaluation Report dates,

- Purchase Order or Subcontract issuance date,

- fabrication progress,

- actual dates of shipping,

- date of arrival at destination port,

- Customs clearance date(s), and

- actual delivered to Site date(s).

It should be noted that many of the above items are in evidence in the list shown under Section 6.8.4.6 in respect of a Maturity Management System. That is why I consider that implementation of such a tracking system would not add a great deal of extra work and yet deliver tremendous benefit for the Project Controls Department's personnel. It would also provide much needed comfort for the Corporate Management Team and the Board of Directors.

6.9.8   Effective Use of Expediters

Expediters are not just useful but very necessary personnel when there are many items to be purchased. Their role it is to keep track of and chase up on the progress of the Purchase Orders that have been let, in an effort to ensure that the goods arrive to the Site on time. The Procurement Manager needs to ensure that an adequate number of competent personnel are brought in to take over this task, thereby freeing up the Buyers to enable them to concentrate on getting out the remaining Purchase Orders in good time. Ideally, the Expediters should not be the responsibility of the Procurement Manager but should operate from within a separate Department (or sub-Department) set up specifically for dealing with Expediting/Logistics under its own Manager (or Sub-Manager). In this way, the Procurement Manager would be free to concentrate on attending to all the work necessary to issue the Purchase Orders and Subcontracts.

6.9.9   Avoiding Delays from Employer's Side

All too often on EPC Projects, the Employer's Team tries to involve itself heavily in the Procurement process. This is in the mistaken belief (in my opinion) that somehow the end product will be inferior if the Employer's people are not allowed to get involved. Unfortunately, I have noticed that such involvement was generally regarded as inter-ference by the Contractor's personnel, which tended to legislate against cooperative working. That attitude was somewhat understandable, because I observed many times that, even though it many have been inadvertent, the Employer's Team ended up dis-rupting the Procurement process and causing delays for the Contractor, all without any noticeable benefits accruing for the Project. The Employer's Team does have legitimate concerns that need to be addressed. However, that can be achieved by the Procurement

Department taking the following steps which, if done properly, should be enough to persuade the Employer's Team to relax their grip a little on what is, after all, ultimately the Contractor's responsibility to get right, not the Employer's responsibility:

1. Conducting formal 'third party due diligence' appraisals, not only on all Vendors and Subcontractors but also on all those supplying logistics support services (such as transportation and Customs clearance functions). The purpose of this is to ensure that such entities are reputable, and that they employ honesty and integrity in all their business dealings. It is particularly important to establish that they have no previous record of participating in bribery and corruption, or engaging in money laundering.

2. Ensuring that all Procurement Department staff have been thoroughly vetted and not found wanting in regard to dealing honestly and with integrity in business.

3. Sharing the contents of the Materials Requisition with the Employer's Team before being issued to the bidders. On this particular point, I have observed that it is more often than not a mistake to jump the gun and go out to bid early if these documents are not close to being 99% of what is finally required. To do otherwise will most likely end up with far too many bidders' queries being raised and, inevitably, a lot of time being expended in getting the requirements fixed properly so that all bids received are on the same basis. A further point is that the Contract Documents often stipulate that the Employer must approve the technical content of the Purchase Order before it can be issued to the potential Vendor. If that requirement is ignored, it can cause a big problem for the Contractor if there are key points of objection from the Employer's side to the technical content. Further, if the Employer's Team later decides not to issue a system or sub-system completion certificate due to being unhappy with technical aspects of materials, goods or equipment, it will almost certainly cause tremendous problems (including delayed Project completion). This will be especially true if the Employer insists on last-minute rectification/remedial work being carried out before certifying completion.

4. Demonstrating to the Employer's Team that the contractual/commercial 'terms and conditions' for the Purchase Orders and Subcontracts include all the required flow down (back-to-back) clauses that match the Contract provisions between the Contractor and the Employer. Typically, this would cover at least such topics as anti-bribery, anti-corruption and anti-money laundering requirements, audit rights of the Employer, etc.

5. Maintaining strict confidentiality in respect of the bidding process, including controlling and faithfully recording bid openings and communications between the Contractor's Procurement Team and the bidders.

6. Involving the Employer's Team and obtaining agreement in advance whenever it becomes necessary to discuss and agree technical changes with a Vendor or Subcontractor.

I have seen some Project Managers try to prevent the Employer's Team from getting the above level of access and transparency about the Procurement Department's activities; it never ended well.

6.9.10   Due Diligence on Vendors and Subcontractors

The Procurement Department is usually tasked with arranging all the logistical support for delivery of goods and equipment to the Site after delivery at the ports of disembarkation. Especially if the Site is in a foreign country, great care must be taken to ensure that the local people engaged are very familiar with the work to be done, and that they have a clean business reputation. This is very important nowadays, not least due to the need to conduct the 'due diligence' checking of third parties demanded by international anti-corruption and anti-bribery laws. Those laws include, but are far from limited to, the USA's 'Foreign Corrupt Practices Act'[24] and the UK's 'Bribery Act 2010',[25] which have a far-reaching effect, well beyond the shores of the USA and the UK.

Failure to conduct such due diligence checks, regardless of how 'squeaky-clean' the third party may appear to be, is likely to result in a Non-Conformance Report being issued against the Contractor by the Employer. Further, if a long-term logistics support provider is actually found guilty of bribing local officials (for example, to facilitate early release of goods from Customs) it will not be an excuse that due diligence checking was not done by the Contractor because of the long-standing relationship between the parties. This is because the international compliance authorities will consider the Contractor guilty of not having done enough to stop the corruption, and thus the Contractor will be tarred with the same brush as the errant logistics support provider.

6.9.11   In-Country Services Problems

When working in a foreign country, the Contractor is well advised to arrange the road transportation of goods and equipment through reputable local hauliers. Such entities will know the problems involved for transporting over-sized and/or heavy loads (such as distillation columns, gas turbines and the like). They will also already have good connections with the various police departments in each area through which the vehicles must pass. Attempting to save money by utilising the Contractor's own vehicles and drivers would usually be a recipe for disaster. However, the Contractor would still need to make sure that the local companies it is relying on to transport the goods and equipment safely are in fact doing their job properly, since the impounding for a transportation violation of equipment that sits on the Project's Critical Path could prove disastrous to the Contractor.

It is therefore worth the time and money for any Contractor who wishes to work in a foreign country for a long time to employ trustworthy local personnel to work inside the Contractor's own organisation as checkers. They should be made responsible for looking after the Contractor's interests by keeping close tabs on what the local support companies are doing (or should be doing). However, the employment of such personnel does not obviate the need for the Contractor to conduct due diligence vetting of all its agents involved in logistics activities, especially those who are likely to be interfacing with Government/Public Officials on a regular basis. The Contractor must take all reasonable steps available nowadays to check that the proposed agents have not been

---

24 Ibid., Section 6.2.2.
25 Ibid., Section 6.2.2.

found guilty of bribing officials (or others) in order to make things happen quicker than they would otherwise have done (such as for road and/or bridge closure permits to facilitate the movement of wide loads).

6.9.12 Customs Clearance Problems

Customs clearance alone can be a very complicated undertaking in a foreign country, and involves such things as the following:

1. The preparation and submission of documentation required to facilitate export or import of goods (equipment and materials). Very often, this involves obtaining certifications from overseas embassies regarding such things as 'country of origin'.

2. Properly representing the Employer during examination and assessment of the goods by Customs officials.

3. Payment of any duty on behalf of the Employer, or ensuring that all paperwork is complete to obtain what is often known as 'levy exemption' for the Employer.

4. Taking delivery of the goods from the Customs authorities after all clearance procedures have been completed.

One area of Customs clearance that sometimes catches a Contractor out is the extraordinarily long time it can take to obtain special import licences for such things as equipment that is used for communications or that have radioactive components on board (no matter how low the radiation level involved). It is best to enquire about such items very early on, in order to get all the necessary paperwork sorted out. Such paperwork can take many different forms, and the requirements and documents to be supplied can change without warning in some countries. In one recent case I was involved with, it took over 18 months to get the necessary permit to import some essential gas igniters that contained a very small radioactive component. Added to that, it often falls to the Employer to be the signatory of such documents, but it is amazing how many times Employers cannot believe that the Contractor's signature alone will not be adequate; such an impasse can take months to resolve. The principal types of documents involved are as follows:

- End-User Certificates for the export of sensitive goods/equipment to countries that are on international security watch lists (such as, but not limited to, Iraq).

- Pledge letters for the re-exporting of any products containing radioactive materials, or for the importation of goods that may be used for other purposes in the destination country and so may compromise security there (such as where sophisticated communication equipment is being imported).

6.9.13 Reducing the Risks of Non-Performing Service Providers

A number of Contractors may well have long-term arrangements with shippers, but the recent (2016) experience with the Hanjin Shipping Company has shown that it is best to deal with more than just one support company, in order not to be left stranded if one of them gets into difficulties. Having said that, special clauses may need to be negotiated that allow the Contractor to get access to its stranded goods without undue difficulty.

Having more than one shipping company on board, and letting each alternative company know that, also has the advantage that it adds an element of competition into the services. Of course, this philosophy is not just useful for reducing the risks involved in getting goods and equipment shipped; it can also be applied to every other support service provided to the Contractor by third parties.

The Procurement Manager must also be careful not to let any Vendor sublet to countries that may have restrictions in place for exporting goods to the country in which the Site is located. For example, some South East Asian countries may have a policy of not exporting goods produced in their country to certain countries in the Middle East. To overcome this problem, the terms and conditions must clearly state that no subletting must be done to sub-vendors/manufacturers who are based in countries outside the Vendor's country, unless that situation has already been identified and embodied in the documentation that forms the contractual agreement between the Vendor and the Contractor.

6.9.14 Obtaining All Available Discounts

Wherever possible, the Contractor must take advantage of any discounts available for large purchases from Vendors where such preferential arrangements are in place. It is possible to pick these up as 'year-end' adjustments with reputable merchants, but it obviously makes sense for the Contractor to have its own centrally stored quantities data readily available, so that easy checking of the final accounting can be handled. However, there is far more likelihood that discount opportunities will be lost due to small orders being placed as a result of the Material Take-Offs (measurements) being short (under-assessed). I have found that this situation is much more likely to occur where the Material Take-Offs are compiled by the staff of an Engineering Subcontractor.

The best way to overcome this problem is to insist that the Subcontractor's Lead Discipline Engineers are each given the responsibility for identifying where each different material is likely to be used on the Project, even if the Detailed Design will occur some months later. Preliminary 'guesstimate' figures of the likely quantities to be involved can then be calculated for each location/application of those materials. In this way, if it becomes imperative to place the order for any given material before all the final quantities can be assessed, it gives the Contractor the option of placing an order with a provisional allowance added in for any unassessed final quantities, in order not to lose out on possible bulk order discounts.

6.9.15 Arranging On-Site Support Services

In regard to signing the Service Agreement for on-Site support services required from the Vendors, it can often prove far more costly to do so if left until after the associated Purchase Order has been issued. This is especially so if the Site sits inside a country where personal security is at greater risk than in the Vendor's home country. Vendors may well use the trick of trying very hard to avoid signing the on-Site Service Agreement until after the Purchase Order is in place. This pitfall can usually be avoided if agreement to all the terms and conditions for the on-Site Service Agreement are settled well in advance of the date when the Purchase Order signing is expected to take place. This particular situation tends to arise more frequently where Long-Lead Items are involved that have to be installed in war-torn or strife-ridden countries.

Getting agreement from the Vendors in such cases is usually much harder to achieve than one may think. The Vendors are prone to offering excuse after excuse that combine to delay agreement considerably. One such excuse is that its Board of Directors must be seen to take extra steps to ensure the protection of its employees who will be working on the Site. On the surface, that does not sound (and is not) at all unreasonable. However, the reality is that those same issues will almost certainly be very quickly resolved and agreed upon after the signing of the Purchase Order. Not least, that will most likely be because the daily rates for the Vendor's employees will have increased significantly above the level of fair and reasonable pricing. Being forced into agreeing to such extortionate pricing tricks can be avoided with good forward planning and hard work.

6.9.16   Ordering More Spares than Necessary

There is a warning for the Contractor in regard to the purchasing of essential spares. This is that a great deal of the advice as to what items are actually needed as spares comes directly from the Vendors, since they are usually out to make as much revenue from sales as possible. That often then results in the ordering of items that will never be required (adding extra unnecessary costs for the Contractor that will not be reimbursed). It also adds unnecessarily to the list of Spare Parts Interchangeability Reports (SPIRs) to be completed. Getting the SPIRs filled out properly by the Vendors can be an exhausting task, as it seems to be one which the Vendors are reluctant to do unless they are pushed hard to deliver them. Despite many in the Oil & Gas industry having different ideas, most Project Management Consultants make big play out of the importance of receiving all the SPIR forms properly completed.[26] The Contractor must therefore pay attention to getting this work completed early and, in order to ensure that handover will not be delayed unnecessarily, not allow the Vendors to prevaricate.

6.10   The Expediting/Logistics Manager

6.10.1   Primary Responsibilities

Monitoring of the activities of the Expediting Manager and the Logistics Manager very often falls under the Procurement Manager's responsibility. These are generally separate activities on mega Projects, but on smaller Projects the functions may be merged under just one individual; there seem to be no hard and fast rules. For convenience I have considered the latter possibility, and the following sets out the typical primary responsibilities of an Expediting/Logistics Manager:

- Manage, organise, direct and monitor the activities of the personnel within the Expediting/Logistics Team.

- Liaise closely with all Departments to understand their logistics requirements, and ensure good logistics support is provided for all parties.

---

26 Eales, P. (n.d.). *SPIR's: are they worth the paper they're written on?* http://mroinsyte.com/spirs-are-they-worth-the-paper-theyre-written-on (accessed 28 January 2018).

The biggest risk for the Project Manager is that the Expediting/Logistics work is not being handled properly, such that unexpected delays then occur. The Project Manager must therefore insist that the Procurement Manager keeps good tabs on what is going on and reports back promptly if things are not happening as they should in either field of activity. This would then enable the problems to be caught in good time, with little lasting negative impact on the Project's objectives.

6.10.2   The Expediter's Role

It is essential for the Contractor to follow up on the status of the various deliveries under the Purchase Orders. If chasing up is not done as a continual routine operation, Vendors can become complacent. If ships have to be cancelled at the last minute due to the materials, goods or equipment coming out of the factory later than expected, it becomes very costly. Added to that, a new ship booking may not be able to get the required items to the Site at the time needed. This is just one reason why an adequate number of competent Expediters is important, whose job is to take over the administration (chasing up and running) of the Purchase Orders once they have been placed. However, in order to maintain strict financial control, the handling of changes to an order should be passed back to the Procurement Department's Buyers to handle, not simply entrusted to the Expediters.

6.10.3   Where Vendors Can Fail

Another important task for the Expediters to follow up on is the delivery of the Vendor Documents. This is vital, so that the Engineering Team will be able to incorporate all the details from the Vendor's designs onto the Contractor's construction drawings. However, this is another area where the Vendors are often late in supplying the information unless pushed hard. It is therefore important for the Contractor to ensure that there is some form of penalty (or, better still, incentive) built into the Purchase Order provisions to encourage the Vendors to perform. One way of doing that is to include a specific priceable line item in the Request for Quotation for completion of the Vendor Documents or, alternatively, to link a significant percentage of the Purchase Order payment terms, e.g. 20%, to completion of the Vendor Documents. I have observed that the latter approach is best, since many bidders deliberately do not price the line item, although negotiation may be needed to establish the final 'withholding' (retention) percentage to be applied.

Another area where Vendors often fall down is with regard to getting the Shipping Documents prepared both properly and on time. Failure to get such documents correct can hold up importation and lead to the Contractor paying out hefty demurrage and/or storage charges. The Expediters should therefore be made responsible for obtaining preliminary (draft) copies of the Shipping Documents as early as possible, and for checking their completeness and validity.

6.10.4   Spare Parts Documentation

The final point I wish to make about the responsibilities of the Expediter relates to making sure that the Vendors complete the SPIRs properly and in good time. Without the SPIRs being finalised, the Employer's Team will generally consider the Contractor's work to be incomplete. In theory, it should be easy for experienced Vendors to complete the SPIRs, but I have never been on a single Project yet where there were not far too many critical SPIRs either uncompleted or improperly completed when it came to the Contractor claiming that completion had been achieved. Attempting to get outstanding SPIRs from a Vendor who has already been fully paid up is very often like trying to squeeze blood from the proverbial stone. Again, the Procurement Manager must ensure that the provisions of the Purchase Order fully deal with the importance of the completed SPIRs and the penalties that will apply if they are not supplied properly and on time.

6.10.5 The Logistics Support Activities

The activities undertaken by those handling the Logistics support requirements are as follows:

- In conjunction with the Procurement Manager, organising and tracking inbound shipping requirements from overseas destinations. This may include assessing and recommending which the best carriers are, as well as what the optimum shipping routes are.

- Organising the local transportation requirements (usually through local service companies) in order to get materials, goods and equipment to the Site by the best possible route.

- Organising the initial receipt, intermediate storage and distribution of materials, goods and equipment to the Site in a timely manner.

- Ensuring that appropriate warehousing will be available in good time to receive and store sensitive materials, goods and equipment. This may also include the Logistics Manager being responsible for recruiting and training the necessary warehouse personnel, who will be required to log all the deliveries into the warehouse, as well as record the final distribution activities for all such stored items.

The above activities require a great deal of communication to take place between all the parties involved (very often over long distances, locally and internationally). The biggest risk faced on the Project is the failure of the third parties to do their respective parts in ensuring efficient movement of materials, goods and equipment to the Site. When operating in overseas locations, it is therefore essential that the Logistics personnel are equipped with reliable communication devices and have access to good communication networks. That will then allow them to (i) get to know about problems as soon as they arise and (ii) contact other parties that could help to resolve the problems quickly. Not arranging for effective communication could lead to many annoying and unnecessary delays occurring with key deliveries.

6.11 The Construction Manager

6.11.1 Principal Function and Responsibilities

Those people who have gained solid experience as Construction Managers have the ability to participate to good effect in the preparation of the Contractor's Bid Proposals. They can also offer sound advice to others preparing various sections of the Project

Execution Plan (PEP), as well as to those developing the Project Schedule. However, the principal function of the Construction Manager is to manage the day-to-day construction work on the Site and, very often, the actual person to be appointed for the Project is not brought on board until the Detailed Design work has commenced in earnest. Nonetheless, if the expertise of a spare experienced Construction Manager can be brought to bear in the early stages, the Project will most certainly benefit a great deal.

The primary responsibilities attaching to the role of Construction Manager are as follows:

1. Plan and organise both the temporary works and the permanent construction work in line with the agreed PEP and Project Schedule.

2. Advise on the mobilisation requirements for all things, from the building of any Base and/or Fly Camps through to obtaining the necessary construction equipment and manpower for the start-on-Site activities, and monitor the follow-up actions.

3. Coordinate/liaise throughout the execution of the construction work with the Engineering, Procurement, HSE, QA/QC and Commissioning Department Managers, as well as the Employer's Project Management Consultant and the Private Security Company's personnel. The principal objective must be to ensure that the Project's requirements are fully understood by all parties, and that all obligations arising therefrom can and will be satisfied in a timely manner. Such obligations extend to safe working, environmental protection and local community liaison while the construction work is ongoing. All concerns in any of those areas must be faithfully communicated in a timely manner to the Project Manager for his/her assistance.

4. Define clear roles and responsibilities for all the Construction Team management personnel, and organise, direct and monitor their activities, including their handling of the work of the Subcontractors. This must be accompanied by giving clear instructions/direction regarding the scope of work that each Manager is to be responsible for, especially with regard to adherence to the time-line in which each section of work is to be completed.

5. Monitor construction work productivity levels and schedule compliance, and investigate the reasons for any less-than-satisfactory performance observed.

6. Ensure all changes to specifications, work scope and drawings are fully documented.

7. Review and approve/modify the manpower/construction equipment resources forecasts for all major on-Site activities.

8. Organise and monitor the preparation of Task Hazard Identification and Risk Assessment reviews as well as the construction Method Statements, and ensure that systems are in place to obtain 'Permits to Work' before the respective physical work activities are allowed to commence.

9. Check regularly on the HSE team to ensure that there is proper evidence of Tool-box Talks/Training being given in respect of all activities where there could be even the smallest risk to life or limb.

10. Ensure all construction work is carried out safely and in accordance with the approved design drawings, and that the required quality standard is achieved, with avoidance of non-conformances and minimising of Punch Items.

11. Ensure that the progress of construction work is in accordance with the Project Schedule, and organise and undertake accelerated working and catch-up plans to avoid/mitigate delays.

12. Institute measures for improvement in the construction processes by applying modifications to the current operating procedures, derived through discussion with those responsible for managing the work at the respective work fronts.

13. Be aware at all times of new risks that arise, and advise 'upper management' of the problem in a timely manner on each occasion if it is not something that can be adequately handled on Site.

14. Ensure adherence to the safety standards applicable to the Project (usually the Contractor's standards, modified to incorporate any additional requirements of the Employer that are established through conducting a 'gap analysis'). In addition, promote a strong safety culture throughout all levels of workers, from the cleaners and office boys through to the on-Site manual workers, right down to the Commissioning Teams.

15. Ensure that a team of people under a responsible individual is deputed to clear up all rubbish on the Site on a daily basis and also to dispose of it properly.

6.11.2 Working Quickly, Effectively and Safely

The Construction Department's primary focus should be on completing the work as quickly and safely as possible, with the minimum amount of reworking and only the smallest number of Punch Items to clear. Full attention must be paid to ensuring that the health and safety of the workforce is given topmost priority. However, the requirement to do the construction work quickly can sometimes appear to conflict with the requirement to ensure the safety of the workers. The reality is that if just one major accident occurs due to taking a shortcut, the amount of time lost in getting permission to restart the stopped work will lose far more time than was gained by employing that shortcut. Added to that, the 'loss of face' (reputation) that follows is not worth the Contractor taking the risk of not fully implementing all the safety requirements.

6.11.3 Poor Quality Work Delays Pre-Commissioning

The pre-commissioning activities generally form part of the construction work, and any problems that arise while conducting those activities will almost certainly need to be put right by the Construction Manager. The Construction Manager should therefore keep a watchful eye on the quality of the construction work to ensure that the need to undertake reworking is obviated since, otherwise, serious delays to the pre-commissioning work could occur.

6.11.4 Implementation Pitfalls

All of the primary responsibilities listed above have risks attached that, if not dealt with adequately, could negatively impact the chances of the Project being completed successfully, as discussed below.

6.11.4.1 Not Following the Project Schedule

Planning and organising the construction activities is a very detailed undertaking that can sometimes expose weaknesses in the original PEP and/or the Project Baseline Schedule. When such problems arise, there is sometimes a tendency for the on-Site Construction Team to shrug its shoulders and carry on with its own plans without informing anybody else, simply regarding the earlier scheduling 'mistakes' as being 'water under the bridge'. This has a habit of rebounding horribly on the Contractor later down the line when attempting to claim an extension of time for genuinely excusable delays. That is because the Employer's Team will inevitably point out that the construction work has not been following the agreed plan. To avoid this situation, the Construction Manager should discuss any perceived problems with the Project Controls Manager and then jointly notify the Project Manager about the proposed solution, following which the Employer's Team should be involved in agreeing changes to the currently agreed Project Schedule.

6.11.4.2 Not Holding Regular Formal Meetings

Coordination/liaison with the Managers of all the other Departments whose actions or non-action would impact the progress of the construction work can be a time-consuming undertaking. For that reason, it is often conducted on an ad hoc (impromptu) basis, not a formally recorded basis. This is not a good way to keep everybody properly informed, since it can lead to the situation where people essential for implementing certain activities are unaware of the need for their input until it is too late. This can be overcome by holding regular formal meetings on a pre-scheduled basis (preferably weekly), where simple minutes are kept that record what needs to be done, by whom and by when. I have heard some Managers state that such meetings are a waste of time, and then later heard those same people complain that they have not been kept informed enough about what is going on. Human beings have not yet developed telepathic communication, so getting together as a group to discuss key issues is essential and, if properly organised, such meetings need not take up too much time.

6.11.4.3 Not Preparing for Mobilisation Adequately

Mobilisation comprises more than simply turning up on the Site to start work, and many Employers have very clear views as to exactly what needs to be in place before the Contractor can start work on the Site. This is especially true in remote areas where Base Camps and/or Fly Camps must be provided by the Contractor. Most of those requirements are perfectly reasonable and have to do with health, safety and environmental issues. The following are examples of such requirements:

1. It must be demonstrated that the welfare of the workers is being taken care of by virtue of the provision of good quality accommodation and associated facilities, such as toilets/showers, canteens, recreation facilities, etc.

2. It must also be shown that Site orientation training has been given to all workers, and that appropriate Toolbox Talks/Training have been (or will be) given by qualified personnel (preferably the Job Performers) for all tasks to be performed. It is essential too that appropriate Task Risk Assessments are properly conducted beforehand.

3. Construction equipment must be in good condition and not represent a danger to the operators or be likely to pollute the environment. Many Employers insist that the operational worthiness of such equipment must be demonstrated by certification provided by approved testing authorities.

4. Construction equipment operators must evidence proper training by way of recognised certification, and sometimes this applies to car drivers also (especially if they will be required to drive on the Site).

5. Job Performers must have received proper training, and they must have applied for and obtained any necessary Permits to Work for the physical on-Site tasks that need to be undertaken.

6. All necessary static and mobile security services must be in place before any work commences on the Site. *(In this context, static security guards are those assigned to the Site or Camp areas, where they do not move off the area they are assigned to, although they may sometimes form part of security patrols throughout the area. They may even occupy fixed positions, such as entrance gates, where vehicles and people are constantly coming and going. Some may even be assigned to watchtowers around the perimeter of a facility. Mobile security services have nothing at all to do with portable telephones. In the context of overseas Projects in potentially or actual hostile areas, the term mobile security services refers to the secure movement of personnel from one reasonably safe area to another. That is usually done in a small convoy of armoured vehicles, with armed guards on board, where the security personnel travelling in the vehicles are all in radio control with each other. The guards are generally fully trained and will wear full body armour; very often, the passengers will also be required to wear similar body and head protection.)*

All of the above takes time to arrange, and will require the implementers to report regularly to the Construction Manager to let him/her know exactly what the status of the mobilisation effort is for each activity, and whether any unforeseen problems have arisen that need decisions from the Construction Manager to resolve them.

6.11.4.4 Not Making Surprise Work Area Visits

Once the on-Site work commences, the Construction Manager needs to be kept fully informed on a daily basis as to what is not going according to plan, and there are many different views as to how that should be done. However, I have noticed that there seems to be no better way than for the Construction Manager to devote a couple of hours each day visiting key work areas and talking directly to the personnel responsible for managing the different sections of work. It is best not to announce the timing of such visits to any area in advance, and for the Construction Manager simply to pitch up at random times in order to get an appreciation of how each Sub-Manager is performing; this really does help to keep people on their toes. Having said that, each Sub-Manager should have a report ready on the previous day's activities to hand over to the Construction Manager in the event that he/she arrives at his/her respective work areas. Of course, those same reports should be lodged each day at the Construction Manager's office (for permanent record purposes), whether or not the Construction Manager pays any particular work area a visit that day.

6.11.4.5   Not Holding Job Performers Accountable

Although the Construction Manager will be held accountable if safety violations occur on the Site, he/she is very much dependent on each Job Performer obtaining the necessary Permits to Work, conducting appropriate Toolbox Talks/Training and ensuring that all workers are wearing the appropriate Personal Protective Equipment. If any Job Performer is found to be consistently unreliable in carrying out the required tasks, then the Construction Manager should not hesitate to remove that person from the Site and find somebody who can do the job better. A good point to bear in mind here as to when to remove somebody for not performing HSE administration work properly is what some have called 'Goldfinger's dictum' (of James Bond fame, as created by Ian Fleming)[27] – 'Once is happenstance (accident/bad luck), twice is coincidence, the third time it's enemy action'. Or, to put it another way: 'Three strikes, and then you're out!'

6.11.4.6   Not Documenting Changes Adequately

Ensuring that all changes to specifications, work scope and drawings are fully documented often gets overlooked in the haste to get the job done, until somebody realises that the expected income looks likely to be less than the probable expenditure. The Construction Manager cannot be held accountable for any unwarranted extras having been wrongly included in the Issued for Construction drawings and specifications. However, he/she would have to take responsibility for not making sure that the members of the Site-based Cost Management Team keep a good eye open to pick up, properly record and then report all changes to the drawings, specifications and work scope occurring after the on-Site work has commenced. The way to mitigate the chances of things being overlooked is to get the Cost Management Team to sign off every month against a statement saying that: (i) all newly issued drawings, specifications and work scopes for the month have been properly reviewed, and (ii) all material changes that can be claimed as extras have been picked up and will be processed in conjunction with the Contract Administration Team. A list of all such changes should be attached to that statement, of course, so that the Construction Manager can monitor future progress on those issues.

6.11.4.7   Not Limiting Authority for Managing Resources

The Construction Manager must make it clear to all Sub-Managers that he/she is the only person on the Site authorised to make changes to (i) the duration of the on-Site activities and (ii) the manpower and construction equipment resources necessary to complete them. Any Manager requiring such changes must report the matter to the Construction Manager formally and obtain approval of the changes before implementing them. Failure of the Sub-Managers to do this could very easily lead to unrecoverable delays being experienced, as well as adding extra costs where there is no budget available to pay for them. Of course, if additional resources are likely to become necessary to give

---

27 Fleming, I. (1958). *Goldfinger*. Ian Fleming Publications Limited/the Ian Fleming Estate.

effect to such implementation changes, then the Construction Manager must first bring the problem to the Project Manager's attention and get permission before proceeding to change the plans.

6.11.4.8   Not Assessing Productivity Properly

Formally assessing productivity levels is something that needs to be done on a consistent basis by the Cost Management Team. It is not an easy task to do and therefore should generally be restricted to elements of work that are scheduled to take a long time and involve large quantities of work. The purpose is to check that actual progress is matching the planned progress. It can be applied to welding work, piping work, instrumentation, etc., as well as to concreting work, steelwork fabrication and erection, etc. Once the true productivity level is found to vary substantially from what had been expected, steps can then be taken to see what had gone wrong and identify what needs to be done to rectify the problem.

Without such analysis, the problem will not be seen until too late. For example, let us assume that only 15% of the instrumentation has been installed, using up 50% of the originally allocated time allowance for that activity, utilising the same-sized team as was allocated for the whole activity. If that inadequate level of performance was evidenced across a number of different critical activities but not discovered until too late in the day, timely completion of the Project would be in jeopardy. Timeliness of discovering such problems is therefore essential, and it is the Construction Manager's responsibility to ensure that adequate productivity reports are submitted promptly, so that the necessary steps to rectify any problems found can be taken as soon as possible.

6.11.4.9   Not Progressing Completion Certification Adequately

Before the Contractor will be allowed to commission a major piece of equipment or a critical system or sub-system, the Contractor will be required to demonstrate that all required validity checks have been conducted, and that everything necessary is in perfect order. However, it does not stop there. All the documentation that goes into validating each piece of equipment, as well as all the various systems and sub-systems, then needs to be compiled into one consolidated set of completion information. That documentation then has to be submitted to the Employer prior to handover taking place. It is an arduous, time-consuming task that cannot be done overnight; work on this has to commence as soon as possible, and it needs to be attended to constantly, all the way through to Project completion.

Unlike a few years back, many software systems now exist that cater for equipment information and data to be used in a much more comprehensive way than the old spreadsheet-based databases used to allow. Employers therefore now expect Contractors to use such enhanced software so that the Employer can obtain the benefit of all the additional reporting available for operational purposes. The best way for the Contractor to deal with the completion documentation requirements today is therefore to bite the bullet and implement a proprietary computerised Completions Management System (CMS). Doing that will enable the Contractor to demonstrate comparatively easily (compared with the old ways of handling documents and data) that the technical integrity of the completed facility has been tested and verified all the way through all the manufacturing and installation processes.

In short, a worthwhile CMS will operate as an effective management control system for the inspection and test activities associated with items provided by the Vendors, as well as for all the on-Site installation work. However, I know of some Contractors who still avoid using a CMS, due to what is perceived to be the high costs involved in its implementation. My experience is that such systems are expensive initially, but once the bedding in has been overcome and experienced teams are retained to work with them, the cost benefits start to come in for the Contractor. Those benefits arise because of the power of the CMS to store, retrieve and present regurgitated data in very attractive ways, saving a great deal of time over somebody trying to do the same thing manually. This is because the CMS will be populated with a full set of engineering data for each element of work, system and sub-system, including such things as the Data Sheets, etc.

In addition, the CMS's data registers will hold all the references to the areas and item coding for each element of work, its system number, etc., as well as references for inspection and test data/results, images and so on. The CMS can also be employed to reference all other pertinent documents and data, such as the operation and maintenance manuals, specifications, calculations, drawings, Vendor documents, etc. Further, the CMS will be capable of generating all the checklists required as the work progresses, as well as all the various reports required by the Employer's Team.

A good CMS is capable of compiling the necessary completion dossiers for each element of work, including all the supporting documents to provide the evidence that the work has been completed satisfactorily. The minimum requirements for such completion information (certainly in respect of systems and sub-systems) often runs to the following:

- Work element description (work item, system or sub-system),
- Delineation of the work element (suitably marked up on drawings),
- Technical Queries,
- Instrument Interconnection Drawings,
- Cable Schedules,
- Factory Acceptance Test Reports,
- Site Acceptance Test Reports,
- Non-Conformance Reports,
- Modifications List,
- Mechanical Completion Checklist,
- Mechanical Completion Report,
- Mechanical Completion Certificate,
- Pre-commissioning Test Report,

- Punchlists (and proof of clearance of all items therein),

- Red-Line Drawings,

- As-Built Drawings,

- Ready for Commissioning Certificate, and

- Commissioning Report.

Making sure that all the above documentation requirements come together at the right time needs organising; it does not all fall into place by accident. In the days when all the documentation was compiled in hard copy form only, it was essential to have personnel dedicated to compiling the information bit by bit. And when it was finally all together, multiple copies were required by the Employer, not just for future reference purposes but for operational purposes too. On a major Project, it was not unusual to have two different Managers to handle this work: (i) the Take-Over Manager and (ii) the Handover Manager. The Take-Over Manager was responsible for ensuring that all the documentation and data required by the Contractor from its Vendors and Subcontractors was ready and available when required. The Handover Manager was responsible for compiling all the documents and data required by the Employer's Team before handover of the facility (or any of its parts, sub-systems or systems) could take place. Both of those tasks required walkdowns to be organised with the respective teams of people involved. Ensuring that such walkdown activities took place on time and did not require redoing was essential if delays to the Project's completion date were to be avoided.

Nowadays, the volume of paperwork to be handled has reduced considerably, and there is an easier route to getting approvals and signatures on documents (electronically applied), all courtesy of the personal computer and EDMS capabilities. This has resulted in the Take-Over Manager's role almost disappearing, leaving just the Handover Manager. However, I have seen recent attempts made to omit even the Handover Manager's position and, instead, to make one of the Commissioning Team's members responsible for handling both the take-over and handover functions. I have even seen different people being used in the taking over and handing over roles for different systems on the same Project. Suffice it to say that I witnessed many problems occurring due to this supposedly streamlined approach that resulted in horrible delays. The problem stemmed from the non-familiarity of the temporary Managers with the requirements for both taking over and handing over, as well as with how to get the required data out of the EDMS in the required format.

My advice here is that a full-time competent Handover Manager should be appointed on a major Project to ensure that everything comes together as required in a timely manner. If this is done in conjunction with a proprietary CMS, then the whole process will be a lot smoother. Not appointing a dedicated Handover Manager and a CMS could so easily lead to delays that result in Liquidated Damages having to be paid.

It should be noted that the task of compiling handover information is not limited to the Construction phase alone. On the contrary, it is very important that it is progressed right from the commencement of the Engineering work, through Procurement, and right to the end of the Commissioning work. To achieve this well, it is incumbent on the Project Controls Manager to ensure that the Project Information Manager (PIM) is keeping track of the activities of all the Departments in regard to data collection and recording, as well as monitoring the activities of the Handover Manager. The problem of missing inspection and test reports/certificates, Spare Parts Interchangeability Reports and the like often arises very late in the day. But if the PIM and the Handover Manager are both doing their jobs well, then it should be possible to eliminate this avoidable situation.

The purpose of listing the above completion tasks in detail is simply to highlight just how much effort must go into compiling all the information that will be required by the Employer's Team to get to the point where the facility can be handed over to the Employer for operational purposes. However, I have on a number of occasions seen all this work left until far too late in the day, which then meant that many extra people had to be drafted in to see if it could be finalised on time. The worst case I saw required six extra people being brought in for more than a year, where there was no budget for that team. Additionally, the date for providing the completion documentation was missed by about 18 months. Luckily though, the Contractor was saved from having to pay Liquidated Damages because the Employer too had been late on items that impacted the Critical Path by more than that duration. Other Contractors may not be so fortunate if they do not pay attention to progressing the preparation of the completion documentation adequately and in good time.

6.11.5 Subcontracting Risks

The majority of construction Projects will require the services of Subcontractors for undertaking specialist activities (e.g. for such things as heating, ventilation and air conditioning installations, instrumentation and controls installations [such as SCADA], etc.) and, very often, general construction activities as well (such as piling, roadworks, piping, etc.). The risks involved in engaging Subcontractors are very high, since the Contractor is not directly managing the workers of the Subcontractors. It is therefore essential that suitable control mechanisms are built into the Subcontract documentation such as will allow the Contractor to implement effective corrective action if a Subcontractor fails to do what the Subcontract says should be done. The things that need to be covered to deal with the issue of the performance (and non-performance) of the Subcontractor are all matters that must be incorporated into the provisions of the various Subcontracts by the Subcontract Procurement Team. However, by the time the Construction Manager is appointed to administer the construction work it will often be too late for him/her to have any influence on the quality or content of the Subcontract documentation. The Construction Manager will therefore have to live by what is actually written in those documents, not by what he/she wishes had been written.

The principal problem that arises with Subcontractors is that reported information related to their progress is often inaccurate. This can lead to the situation where one Subcontractor has been mobilised for a specific start-on-Site date, but the preceding Subcontractor will not be finished and out of the way before the subsequent Subcontractor arrives. A problem very much linked to this issue is that too much reliance is often placed on just one single Subcontractor for a given work scope, and nothing can be done to increase the production capacity in the short time-frame still remaining. Consideration needs to be given to this issue at the time of bidding, to see if it is possible to split big work packages down so that there is always backup available if one of the Subcontractors is having difficulty in meeting its contractual obligations. However, if this has not been done, then the Construction Manager needs to keep a close watch on the progress of large-sized critical work packages, in order to be able to implement contingency plans in the event that things start going wrong. On the other hand, I am aware of a major Project where splitting a large work scope into manageable portions, with each Subcontractor providing its own scaffolding covering its individual work area, failed to work as intended. The primary problem was that the individual Subcontractors were reluctant to share their scaffolding with other Subcontractors, thereby negatively impacting progress. The remedy applied was to remove the provision of scaffolding from each Subcontractor's scope of work and engage a specialist scaffolding Subcontractor to provide common scaffolding to all personnel on the Site, including the Subcontractors.

Most Subcontracts will clearly state that the Subcontractor must employ its own HSE Officers to guide and monitor the Subcontractor's workforce. Nonetheless, having seen how HSE problems can easily arise with local Subcontractors, I suggest that it is always advisable for the Contractor to engage extra HSE personnel in a similar ratio to that used by the Contractor for its own workforce. In that way, the Contractor will be better able to ensure that essential training for special tasks, as well as adequate daily Toolbox Talks, are conducted by the Subcontractors. Such additional personnel will also be available to check that the safe working policy required by the Contractor is being followed faithfully by the Subcontractors at all times of the working day.

In some countries, the consideration paid to preserving human life is, unfortunately, not always given the highest priority. Special attention must therefore be given to repeatedly reminding local Subcontractors of the need to put the safety of the workers first, and to taking stern action wherever violations of the HSE requirements occur. The most numerous safety violations (not accidents) that I have seen recorded on the Projects I have been involved on are (i) not wearing appropriate Personal Protective Equipment (boots, gloves, eye shields, etc.), and (ii) not wearing safety harnesses (or unhitching them) when working at height. Persons guilty of repeated safety offences must be permanently banned from the Site; anything less will not send the right message to the other workers. I personally consider that the three-strike rule might be too generous/dangerous to be applied where such blatant HSE violations occur, so I suggest that a two-strike rule should be rigorously enforced instead. However, because of the difficulties and costs involved in getting workers to an overseas Site, I cannot see many Project Managers adopting that suggestion for the foreign workforce.

Quality of the completed work is another area where Subcontractors can cause problems for the Contractor, particularly if the list of Punch Items for corrective work becomes too large. Poor quality work can often unexpectedly delay Mechanical Completion and thereby also hold up the commencement of the commissioning work. For this reason, it is also advisable for the Contractor to ensure that adequate numbers of Quality Control (QC) personnel are engaged by the Contractor, even in situations where it appears that the Subcontractor has an acceptable QC set-up. This would add extra eyes to be able to pick up quality issues early, and thereby aid in avoiding too much reworking. Sadly, however, many Contractors do not see this as a necessity until too late in the day, by which time the damage has already been done.

6.11.6   Permit to Work Compliance

Some Projects are more dangerous than others, and any Projects where hydrocarbons will be introduced at the commissioning stage rank amongst the most threatening. Certain general activities too are ranked as being very hazardous, such as working in confined spaces and/or in deep excavations, and lifting of very heavy loads. Wherever construction activities carry heavier risks (even if it is from external sources, such as the possibility of an armed attack occurring), there will almost certainly be a Permit-to-Work (PTW) system in operation that absolutely must be followed by the Contractor. In such situations, consistent failure to follow the PTW system will most likely be considered to be a material breach of the Contract, with the possibility of leading to termination of the Contractor's employment. It is therefore imperative that the Construction Manager surrounds him/herself with competent Job Performers who thoroughly understand the PTW requirements, and who carry the responsibility for ensuring that no work is commenced until the requisite PTW has been obtained.

6.11.7   Delay Risks Caused by Suppliers

The Construction Manager is, to a great extent, dependent on supplies/deliveries emanating from outside the Site arriving on time. This situation needs to be monitored to make sure that, if things are not going as planned, then there will be time to put alternative arrangements in place. The Construction Manager cannot be expected to be the one to monitor this, since he/she will have many other things to concentrate on. This means that the right people must be given the responsibility to keep a keen eye on the situation and report to the Construction Manager when things are going wrong. The best people to do that are the topmost Site-based representatives from the Procurement Department, namely the person in charge of expediting and tracking deliveries to the ports of entry and also the person in charge of organising the logistics support activities (for such things as Customs clearance and local transportation). Between them, these two people should know better than anybody else the true status of the supplies/delivery situation for every item, and they should deliver a daily written report on the current situation to the Construction Manager. Not only that, at least once each week, at a pre-arranged time, formal discussions should be held by those personnel with the Construction Manager to talk through the issues regarding each major delivery. This will ensure that everybody fully understands the risks involved in those deliveries not happening as planned, and what action needs to be taken to reduce those risks.

6.11.8   Controlling Wastage

Turning now to another important matter: an area that is very difficult to control on a Project unless a good monitoring system is in place is the management of wastage. This is because excessive wastage of manpower and construction equipment resources, as well as materials, can occur with almost every activity on the Site. Even with a good monitoring system in place, reducing wastage of materials will become less and less important to the workers as pressure mounts to complete tasks quicker. The Construction Manager therefore needs to be seen to be very strict about minimising wastage by insisting that worthwhile KPIs are in place to enable monitoring of the wastage situation to be carried out adequately, and which are effective in quickly identifying where things are going awry.

As mentioned above, wastage comes in two forms:

(i) the invisible wastage of valuable manpower and construction equipment, which can be caused by ineffective working, progress hold-ups and lack of attention about improving productivity (or even meeting the original productivity targets anticipated at the bidding stage); and

(ii) the very visible wastage of materials that occurs daily on a building Site, not to mention the large amount of wastage that can occur through myriad acts of pilferage.

It is not too difficult to come up with a worthwhile list of KPIs to use for comparing actual manpower and construction equipment outputs with what had been expected. Any major difference between the figures should then be checked to see if abnormal circumstances caused the problem or if it is a continuing trend that requires a permanent cure to be found. For materials, the financial leakage caused by excess wastage alone can be enough to wipe out the anticipated profit for some Projects. Reducing the amount of materials wastage effectively is therefore a necessity, not a luxury, and comprises the following four-step process:

1. The work crews and their supervisors need to be reminded regularly that wastage must be reduced as much as possible, and that even the materials for temporary construction work must be re-used wherever feasible (consistent with the cost savings envisaged versus the anticipated costs in reworking the materials).

2. The gross quantities of the materials used for each item of work (including an appropriate allowance for wastage) must be computed in advance of the work commencing, as too must the total number of man-hours that are to be allocated for completing each item of work.

3. The actual quantities of materials used to complete each item of work, as well as the total number of man-hours actually expended, must be faithfully recorded, and the outputs then compared with the original expectations. This should be done at intermediate stages, preferably as soon as practically possible after a particular stage has been reached, as well as at the end of the activity.

4. Where an analysis of the original expectations and the actual amounts used show a big variance, then appropriate steps can be taken to redress the situation before it is too late to do anything about it. Sometimes, further investigation will even throw up the likelihood that large-scale, organised pilferage is going on, and this is why timely analysis is vital. I once saw this occur with paint deliveries on a massive Design-Build housing scheme, where all the estimated quantity of paint had been used up, but less than 50% of the units had been painted.

Without adequate numbers of Quantity Surveyors and/or Cost Engineers to monitor the situation properly, the wastage problem will not be handled effectively. The additional cost of an adequately-sized team that could spare the little extra time that it would take to be able to monitor the wastage situation properly would be very small. I therefore suggest that saving money by not providing an adequately-sized measuring/costing team to help control wastage would be false economy on any major Project.

6.12   The Commissioning Manager

6.12.1   Primary Responsibilities

The following sets out the primary responsibilities of the Commissioning Manager:

1. Manage, organise, direct and monitor the activities of the personnel within the Commissioning Team.

2. Plan, organise and control all commissioning activities, and ensure that all necessary materials and spare parts will be readily available for commissioning purposes.

3. Ensure that all necessary checklists and record forms for commissioning purposes are available for use in good time and are properly utilised.

4. Ensure that all necessary pre-commissioning activities have been satisfactorily concluded before proceeding to implement commissioning activities.

5. Ensure that all documentation that needs to be in place before the Employer's representatives will sign off systems and sub-systems as having been satisfactorily mechanically completed is actually being put in place by the Contractor's Handover Team in the required format. This requires a systematic approach to completion of that documentation being followed that matches the Commissioning Team's schedule of commissioning activities.

6. Ensure that all precautions necessary to ensure safe commissioning are fully understood by all members of the Commissioning Team, and that adequate safety arrangements are in place. If hydrocarbons are being introduced into the facility, this will most likely require having an ambulance on adjacent standby with a competent medical team on board, and even a large capacity fire engine, complete with attendant water bowser and a full complement of qualified fire fighters.

7. Ensure that all the necessary permits, authorities and clearances have been obtained before commissioning work commences.

8. Arrange for the compilation of all required commissioning reports.

For the Commissioning Manager, there is always the risk that (i) the construction work will not have been completed properly, (ii) the pre-commissioning work will be finished later than required, and (iii) essential handover documentation due from the Construction Team will be missing. Although all such issues are due to the shortcomings of the earlier participants on the Project (principally, the Engineering, Procurement and Contruction Teams), the consequences of any knock-on delays experienced become the problem of the Commissioning Manager. Sadly, all too often the Commissioning Team is squeezed for time in this way due to the actions or inactions of others.

6.12.2 Major Risks

The following sets out some of the major risks that the Commissioning Manager will be responsible for handling and resolving, as well as the best ways to reduce or eliminate those risks.

6.12.2.1 Spare Parts Availability

Necessary commissioning spare parts may not be available on time or even available at all. The way to reduce this risk is for the Commissioning Manager to:

(i) get involved in the procurement process by being part of the team that decides on the spare parts that are required for commissioning purposes, and

(ii) follow up on a regular basis to ensure that the commissioning spare parts are ordered and are being delivered in good time.

Once the commissioning spare parts have been delivered to the Site, the Commissioning Manager should also ensure that they are stored in a secure warehouse (with adequate air conditioning and humidity control to ensure proper preservation of any sensitive goods). All spare parts should also be properly labelled and placed in meaningful lots, so that they can be easily identified later and accessed quickly, as and when required.

6.12.2.2 Commissioning Safely

All commissioning work carries great risks for the safety of humans in the vicinity of the commissioning activities. The involvement of adequate numbers of competent HSE personnel and the conducting of Toolbox Talks/Training is therefore absolutely crucial to eliminating or reducing the risks of something untoward happening. The Commissioning Manager's obvious involvement by active participation in HSE activities and presence at the Toolbox Talks/Training sessions will help reinforce how importantly the Contractor's Corporate Management Team treats the subject of safety at work. This will also have the knock-on benefit of encouraging the workers to take safety matters equally as seriously.

If the completed Project incorporates process work involving hydrocarbons, then there is the very real risk of an explosion occurring if all possible steps have not been taken to ensure that all sub-systems, including piping work, have been completely tested and found not to have any leaks. It is therefore absolutely essential under such circumstances that the Commissioning Manager institutes checks in respect of each section of work in a commissioning activity, in order to establish that all required testing and remedial/rectification work has been satisfactorily completed before any commissioning work is allowed to commence. All aspects need to be checked thoroughly.

For example, where pressure vessels are concerned, if significant torque loss occurs for some securing bolts, the internal pressure when being tested may exceed the compressive force holding the secured component in place. In such a case, a blow-out could occur that causes serious injury or even loss of life for somebody in the path of a flying inspection door or similar object. I was on the Site when a blow-out caused a blind flange on a pressure vessel under test to rip free from its retainers and crash through an adjacent access staircase, depositing itself and a section of balusters and handrailing 25 m away; fortunately, nobody was hurt.

6.12.2.3   Commissioning Equipment Jointly with the Employer's Team

Ideally, all commissioning of essential equipment should be done with representatives from the Employer's Team present to witness what is happening. It would of course be best for the Contractor if the Employer's Team were to assume responsibility for signing off before any essential equipment is put into operation, and some Employers in fact insist that is what must happen. However, some other Employers prefer to stand well back from involvement in commissioning activities, leaving all the risks with the Contractor. It would therefore be a good idea for the Commissioning Manager to set up meetings with the Employer's Operational Team well in advance of the pre-commissioning work commencing to discuss this issue in depth. The objective would be to see how much responsibility for commencing the operation of essential equipment could be shared with or moved over to the Employer's Team.

6.12.3   Emergency Vehicles Requirements

A big surprise to some Contractors is that emergency vehicles are often needed to be present when the commissioning work involves hydrocarbons and, as mentioned earlier, it can mean that the Contractor will need to supply mobile firefighting facilities as well as an ambulance and medical staff. Such facilities cannot be mobilised quickly, and they cost a lot of money to organise. This therefore poses a risk to the Project's profitability if such items were not priced for in the bid submission. In remote locations, there may not even be such equipment available for hire at any price; early checking of the realities of the situation is therefore essential. I was on one Project where the local fire station was too far away, and the Chief Fire Officer refused to get involved in case an emergency call arose while their only fire engine was on standby at our Site. The day was only saved by the unexpectedly early arrival of the fire engine and other firefighting equipment that the Employer had purchased for use in the operational phase of the facility.

6.13   The Operational Readiness Manager

6.13.1   Primary Function

The Operational Readiness Manager may often be a member of the Commissioning Team, reporting to the Commissioning Manager. The primary function of the Operational Readiness Manager is to ensure that all the material necessary to bridge the knowledge gap that would otherwise exist between the Construction Team and the Operations Team in respect of a new capital Project will be made available to the Operations Team in good time (i.e. preferably before the commissioning phase is commenced).

I have seen it written in job specifications that the Operational Readiness Manager must make sure that all plant and equipment will be ready and fully functional on time. Such a requirement is patent nonsense in my opinion, since the Operational Readiness Manager has no control or authority over any of the work elements that would lead to successful completion of the Project (namely, Engineering, Procurement, Construction and Commissioning).

Added to that, very often the Operational Readiness Manager is appointed late in the day and is very busy playing catch-up with his/her own direct workload. Even then, he/she is sometimes called upon to support the Information Management Department in order to help (often on a 'hands-on' basis) close out items that have become delayed (such as dealing with equipment tagging problems). The Operational Readiness Manager should be allowed to function unhindered and concentrate on doing what is an essential role on any major Project, and especially so if the appointment is not made early enough. This is because the Employer will most certainly not deem the Project to be complete if its Operations Team is lacking essential operational or maintenance information.

6.13.2   Reliance on the Information Management Team

There will be a considerable amount of operational and maintenance documentation delivered from the various equipment suppliers. Very often, it will not be in a format that is entirely consistent with, and readily transferable into, the Enterprise Asset Management System (EAMS) that has been selected for the Project (such as IBM's Maximo, for example). In addition, the Contractor's own Engineering Team will have produced many drawings containing the tag numbers for the equipment items (very often numbering many tens of thousands). The very real possibility therefore exists that there will be a considerable number of wrong, duplicated and missing tag numbers, all of which need to be correctly logged into the EAMS. Further, there will be a wealth of spare parts, for which there will be corresponding Spare Parts Interchangeability Records (SPIRs), the data from which must also be faithfully entered into the EAMS. All this should be taken care of by the Information Management Department. However, the Operational Readiness Manager will still need to check that all such information is being correctly logged into the EAMS in a timely fashion, so that there are no untoward surprises later regarding delays to this work.

6.13.3   Inputs from the Operational Readiness Manager

The Information Management Department will only be taking care of logging the details of the components of the as-built facility, and will not be concerned with the maintenance strategies for the completed facility, as that task falls to the Operational Readiness Manager. Operational Readiness prepares the custodians of a facility, as well as their support staff, to be fully ready to assume ownership of the asset at the point of handover. In order to achieve that, the Operational Readiness Manager must:

(i)   identify the systems required for successful operation and maintenance of the facility;

(ii)   develop safe operating and maintenance procedures;

(iii)   ensure that training is given to administrative personnel on new systems and procedures;

(iv)   ensure that training is given to all operating and maintenance personnel for the new equipment, systems and subsystems; and

(v)   ensure that the staff allocated to operate the completed facility will be able to do so safely, effectively and efficiently (which is not the same thing as ensuring that training is given).

6.13.4 Lack of Operational Readiness Information from Others

The biggest risk for the Operational Readiness Manager is that there is insufficient information provided to enable the Operating and Maintenance Procedures to be completed on time. Sometimes this is due to the bidding documents not having stipulated clearly enough what documentation is required from Vendors and Subcontractors. It is therefore imperative that the Operational Readiness Manager reviews the Vendor Deliverables requirements as soon as possible after his/her appointment (assuming that he/she did not have the opportunity to provide input to the Request for Quotation or Invitation to Bid documents). The objective of that review must be to see if there are any potential documentation problems coming up that can be averted through timely corrective action.

6.13.5 Operational Readiness Documentation Focus

There will inevitably be a notable difference in the way Operational Readiness is treated by the Contractor, according to whether or not the Contractor is to be responsible for undertaking the Operation and Maintenance work after the facility has been completed. Where the Employer will be responsible for all the Operation and Maintenance work, the Contractor's sole aim will be to complete the Project on time and within budget. Very often that results in the Contractor doing the minimum necessary to compile the information database it is considered will be required after handover (i.e. only a sufficing effort is employed). This approach brings with it the risk that critical information relating to the assets is nowhere to be found when needed.

On the other hand, where the Contractor will be responsible for undertaking the Operation and Maintenance work after completion of the facility, a great deal more effort will be made by the Contractor to ensure that the documentation/information developed during the Project's Engineering, Construction and Commissioning phases is properly translated into the format that is essential for operational purposes. In addition, whether or not the Contractor adds in additional facilities that will aid in reducing operations and maintenance costs hinges on the Contractor's role in the Operation and Maintenance work. The Operational Readiness Manager must keep this differentiation in mind and check that what will be handed over to the Employer is in line with the Contractor's actual role on the Project following such handover.

6.13.6 Employer Pressure to Improve Facilities

The problem for the Contractor occurs when the Employer's Operation and Maintenance Team gets involved early on and attempts to influence the Contractor's Engineering and Procurement Teams to provide an enhanced facility that the Contractor has not included allowance for in the bid pricing. Strictly from an operational standpoint, the Employer's Operation and Maintenance Team may well be right that the additional facilities requested are highly desirable, but there will be problems brought about by the division between the financial management of capital expenditure (CAPEX) and operating expenditure (OPEX). The gap between the two budgets is not something the Contractor should have to pay for, and the Operational Readiness Manager must pay close attention to what the Contractor should be prepared to pay for, bearing in mind the role the Contractor has to play (if anything at all) in the Operation and Maintenance work.

6.14   The QA/QC Manager

6.14.1   Quality Assurance Versus Quality Control

I have observed that many people who happily talk about Quality Assurance (QA) and Quality Control (QC) do not actually realise that the QA function is quite different from the QC function. The QA function involves setting up the procedures/processes aimed at controlling the quality of the outputs and, subsequently, the auditing of how the work was done to (check whether or not the required procedures/processes were followed properly). On the other hand, the QC function involves checking the quality of the actual work outputs to see if they satisfactorily conform with the expectations set out in the quality procedures/processes, and it also includes witnessing factory and other acceptance tests.

I also saw that the role of the QA/QC Manager was sometimes misunderstood. On several occasions I witnessed conflict within the Project Management Teams due to some Managers expecting the QA/QC Manager to assume the responsibility of managing the quality risks evidenced within the Departments. I even heard an Engineering Manager berating the QA/QC Manager for not allocating QC people to spend time to try to find the equipment 'tagging' errors before the tag numbers were committed to the drawings (and which is, of course, the responsibility of the Engineering Manager to organise, in conjunction with the Project Information Manager).

6.14.2   QA/QC Manager's Primary Responsibilities

The following sets out the primary responsibilities within the QA/QC Manager's overall job function of managing the QA/QC Department and organising, directing and monitoring the activities of its personnel:

1. Establish and implement the Project's Quality Management System, and ensure that each Department is made aware of the contribution required from it to achieve the quality standard specified for the Project.

2. Assisted by QA/QC Engineers, develop the Project's Quality Plan and the Inspection and Test Plans, and ensure that all other required procedures, plans and manuals necessary for guidance and control of the implementation work, as well as for conveying essential information, are properly prepared in a timely manner.

3. Collaborate with third-party inspectors for monitoring and expediting fabrication work and the like in the factories of specialist equipment and goods manufacturers.

4. Allocate QC personnel or appropriate third-party inspectors to attend Factory Acceptance Tests, Site Acceptance Tests and the like.

5. Allocate QC personnel to conduct on-Site quality checking of the physical work being undertaken.

6. Carry out regular checks on the QC activities to ensure that they are being conducted effectively (usually by reviewing the QC reports and monitoring the relevant Key Performance Indicators).

7. Carry out regular quality audits on each Department to ensure that the QA/QC procedures/processes are being implemented properly.

8. Ensure that all documentation required for record purposes is being provided and properly handled.

6.14.3   Main Risks for QA/QC Manager

The main risks for the QA/QC Manager are as follows:

1. The processes/procedures agreed with the Employer are not followed by the concerned Department, and the Employer therefore threatens to issue Non-Conformance Reports (NCRs) that will sometimes carry severe financial penalties for the Contractor.

2. The products from the Vendors' factories do not satisfy the requirements stipulated in the Materials Requisitions.

3. The quality of the physical work outputs is poor, leading to substantial items of remedial work being required that the Employer will not accept as being Punch Items, resulting in the contractually required completion date being delayed and Liquidated Damages being applied by the Employer.

6.14.4   Avoiding Quality Assurance Risks

To avoid the QA risks developing into serious issues that could lead to NCRs being received from the Employer's Team, regular internal 'desktop' audits must be conducted by the QA personnel to see if things are on track or going wrong before such matters are picked up by the Employer's Team. There is no shortcut available to the QA/QC Manager to overcome the above risks; it requires hard work. This is because there are many different Departments involved, and so it can be almost a fulltime exercise on a large Project. If this exercise is done properly to ensure that the required processes and procedures are followed, then the benefits of that should show themselves via the good quality of work being delivered.

6.14.5   Reducing Quality Control Risks on the Site

To reduce the QC risks on the Site, it requires a team of dedicated Quality Control Supervisors moving around the Site all day long, on a continuous basis. Their role is twofold:

(i) to dissuade the workers from undertaking the work in a shoddy fashion (which is achieved better when the presence of the QC personnel is more evident), and

(ii) to attempt to pick up as many of what could later become Punch Items, in order that they can be remedied (closed out) before the final walkdown in the presence of the Employer's Team members for completion certification purposes.

However, ensuring that the QC personnel are carrying out their supervision roles properly is not so easy on a big Site, and especially in situations where the weather conditions are uncomfortable but there are plenty of places to take cover from the elements. One way around this problem is to insist that each of the individual QC Supervisors submits a written daily report identifying what the main problems found the previous day were. To add some element of competition into this, a small monthly incentive bonus could be given to the person considered responsible for contributing the most to avoiding expensive rework for the Contractor.

6.14.6 Reducing Vendor Quality Control Risks

To reduce the QC risks in respect of the Vendors' products, the QA/QC Manager must ensure that suitably qualified personnel are engaged to visit the Vendors' factories and witness the inspections and tests identified in the respective contractual documentation. Such inspectors must be charged with faithfully reporting back their findings to the Contractor, as well as being responsible for checking and signing off all Inspection Release Notifications before allowing the materials, goods or equipment to be shipped. Because of the high costs involved in putting the Contractor's own personnel into overseas countries, this work is frequently sublet to a reputable third-party Service Provider who has been previously tried and tested in undertaking such work for the Contractor. This can be an expensive undertaking if left to the last minute. It is therefore advisable for the Contractor to set up long-term Subcontract Framework Agreements with several reputable companies so that the applicable unit rates are fixed in advance and personnel can be made available at short notice. This will then avoid the need for the Contractor to pay excessive premiums each time an unexpected overseas inspection is found to be required.

6.15   The HSE Manager

6.15.1   Importance of HSE Inputs from Everybody

I learnt exactly how important safety matters are when my father was invalided out of the construction industry through his own doing when I was just 11 years old. The repercussions for my mother and her five children were substantial and continued for decades. The point I wish to make here is that I consider safety at work to be the responsibility of everybody, not just the HSE officers. All individuals must contribute to ensuring that not just their own work but the work of all others around them is being conducted safely.

Further, if a worker violates the safety rules, which then results in an accident of any magnitude whatsoever, the responsibility for that accident must first be put squarely on the shoulders of the workers involved, not automatically transferred to the HSE staff. Of course, if the HSE staff had been complicit in allowing unsafe work practices to take place, that is another matter entirely. Generally however, no one person can be expected to be in a position to watch the activities of the 30–50 workers under his/her watch and stop them from doing something silly when it comes to following the HSE rules. Many transgressions are deliberately undertaken at the point when the HSE Officer's eyes are averted elsewhere. Exactly the same goes for environmental matters too. Many HSE problems could be nipped in the bud if everybody realised that their individual input is vital, such as by speaking up on HSE matters whenever something silly is about to take place.

6.15.2   Primary Responsibilities

The HSE Manager is required to ensure that adequate monitoring of the construction work activities is done as a continuous endeavour. The objective is to ensure that safe working practices are being implemented without fail. The following are the primary responsibilities of the HSE Manager:

1. Promote and encourage a high level of HSE awareness, not just at the Site but throughout all locations where the Project staff are working (such as in the Project's off-Site Engineering and Procurement Departments).

2. Manage the HSE Department and direct/monitor its team members in all aspects. One area that is often overlooked by the HSE team is Site tidiness/cleanliness, and I have only ever been associated with just one Project where the Contractor was commended for keeping an orderly Site. Too many times the Employer's Team, quite rightly, complained of a dangerously ill-kept Site, with rubbish strewn everywhere, and which could have caused personal injury as well as health problems. If the HSE Manager does not stress the importance of Site tidiness and cleanliness, it will almost certainly be overlooked. Making Site housekeeping an HSE issue elevates its importance, which then means it stands more chance of being done.

3. Coordinate with Managers in other Departments to ensure that HSE issues have been fully considered in the execution of the Project in respect of the specific activities to be undertaken within each individual Department.

4. Ensure that Task Hazard Identification and Task Risk Assessment reviews for each discrete physical activity have been undertaken before commencement of the physical work, and ensure that all construction work is carried out with adequate HSE supervision.

5. Pay attention to and ensure implementation of any 'golden rules' or similar safety guidelines followed by the Employer's Team.

6. Resolve HSE issues arising from the Project's implementation activities.

7. Continually monitor HSE performance on the Project, and provide monthly statistical analysis reports thereon to the Project Controls Manager and the Project Manager (including findings of safety incident investigations, Lessons Learnt, etc.).

8. Ensure regular HSE audits are conducted regarding the level of compliance of the workforce with the HSE management requirements.

6.15.3   Safety Provisions and Training

Major EPC Projects sometimes have thousands of workers on the Site, often scattered over a wide area and working on a variety of different activities. Each individual will be making hundreds of decisions as he/she goes about undertaking his/her allotted tasks and moving about the Site where others are working.

Sometimes, the HSE Manager is the one first blamed when a worker fails to follow the safety rules and something untoward happens as a consequence. However, provided that the HSE Manager had taken all reasonable steps to ensure that all safety provisions were in place and that all precautions had been taken before work commenced, then I have formed the opinion that no direct blame should be attached to the HSE Manager or any of the HSE support staff. Those provisions and precautions must include the provision of safety barriers wherever needed, Personal Protective Equipment, etc., as well as ensuring that the workers are fully qualified to do the work and have been properly guided before starting (such as via appropriate Toolbox Talks/Training). On the other hand, if all necessary precautions were not in place or suitable training had not been given to the workers, then I would most likely take the view that the relevant HSE personnel were equally culpable for an untoward incident occurring.

6.15.4 Enforcing HSE Rules to Prevent Accidents

Throughout my career in the construction industry I have been close to a number of serious HSE incidents that occurred because HSE procedures were not followed properly by the workers. I have also had friends in the construction industry die because of accidents that should never have happened. Such incidents, non-fatal and otherwise, include the following:

1. Not checking that the diesel tanks had been fully filled with sand before proceeding to cut them up using an oxy-acetylene torch (my Bristolian friends, a father and son, were killed by the blast).

2. A friend accidentally backing into a lift shaft and losing his life when helping to move some mobile scaffolding, where the barrier had been temporarily moved for some adjustments to be made. The lookout man had taken a five-minute break without informing anybody (in order to make urgent use of the toilet facilities) and without attempting to temporarily barricade the area off in any way.

3. Not checking the repacking of a container that had been inspected at the main gate, which then was the cause of the mobile crane being flipped onto its back when it was attempting to lift the supposed 20-ton container over a dyke. The container had been refilled only at the back end (i.e. the load was not evenly distributed along the length of the container), with a further 10 tons of goods added that should never have been there in the first place.

4. Not providing adequate concrete bases for a tower crane sitting in a deep basement in the rainy season, where the flood water caused by an unexpected storm washed all the soil away from the crane's bases due to inadequate dewatering facilities having been provided. It caused the crane to collapse in the middle of a city centre redevelopment Site, where I was sitting just 60 m or so away watching as it came crashing down. Four very unfortunate workers lost their lives that day, due to their inability to get out from under the collapsed crane and reach the ill-placed escape route quickly enough as the water level rose swiftly.

Added to those incidents, I have been involved on too many Projects where deaths occurred through electrocution, due to such things as failure to isolate the power supply properly or running bare wires across the ground. On one occasion, although the cables lying on the ground had been barricaded off, a death was caused about 50 m away after a huge downpour when a worker walked through the expansive puddle in which that cable was then sitting. All those deaths were preventable had the HSE procedures and precautions been followed fully/properly.

My point in particularly mentioning my own experiences is simple: I am far from alone. Many of my colleagues in the construction industry also have first-hand experience of fatal on-Site accidents/incidents. That all adds up to a large number of unnecessary deaths. We therefore all need to go the extra mile to help make the construction industry a safer place. Every single day millions of construction operatives across the world work in what is a very dangerous job, simply to be able to put food on the table for their families. We all need to do our bit to make sure that the work environment is as safe as it can be, and that those workers will return home safely to their families. However, as Briggs and McCabe found, a lot still needs to be done to improve the safety situation for construction workers[28]

6.15.5   Motivating On-Site HSE Officers

The major risk that the HSE Manager faces nowadays is that his/her own supporting staff are not present on the Site as much as they should be. They may spend too much time sitting in the office engaged on 'paperwork', and they may also be too lenient with transgressing workers. Worse, they may not ensure that essential Toolbox Talks/Training are given and that 'readiness checking' is conducted for critical activities before the concerned work commences. The best way to reduce the level of this risk is to require all HSE supervisory staff to hand in a brief written report at the end of each working day to record what work activities were scrutinised (and where) and what violations were observed. This has the effect of turning the HSE monitoring duties into a competition among the HSE Officers, and that in turn helps to keep the workers on their toes and try harder not to let HSE violations occur. As I suggested for the QC Supervisors, offering a small monthly incentive bonus to the HSE Supervisors to do their monitoring job well would also probably help significantly to reduce the number of untoward incidents occurring.

6.15.6   Off-Site HSE Incidents

Where the activities of the off-Site Departments are concerned, the reality is that each individual Department Manager faces the very real risk that their staff may do something that is dangerous to themselves and possibly to others too. This can run from minor things such as people stacking too much on their desks so that the workspace becomes ergonomically unsound (thereby risking long-term muscular and other bodily injuries that could have been avoided), through to serious overloading of power points that could cause a fire. The HSE Manager cannot be held responsible for finding and stopping all such infractions, and it must therefore be the responsibility of each and every one of the Department Managers to monitor and control bad HSE behaviour in their specific work areas, and make sure that suitable HSE awareness training is provided to the Department's personnel. Of course, all such safety problems found should then be relayed to the HSE Manager, so that they can be recorded in the Lessons Learnt register.

---

28 Briggs, L. and McCabe, M. (2012). *Getting Home Safely – Inquiry into Compliance with Work Health and Safety Requirements in the ACT's [Australian Capital Territory] Construction Industry*. Available at www.cdn.justice.act.gov.au

6.15.7   Safety Moment Chats

A very good way to make everybody aware of possible HSE dangers, and thereby reduce the level/incidence of the risks involved, is to insist that all scheduled meetings should be started off with a 'Safety Moment' chat. This could be given by a different person at each meeting, on a rotational basis. Ideally, the topics should deal with a safety transgression spotted (and corrected) since the previous meeting. Only five minutes maximum should be allocated to convey both the problem and the solution to all others in the meeting. This mechanism has the effect of making people more aware as to what is going on around them. Every little improvement suggested can thereby help to improve the working environment and the safety situation, to everybody's benefit.

It is also a good idea to give the general staff a short talk now and then about where things have been found to go wrong, even when simple tasks are involved. This could take place just once a month or immediately after a particular HSE transgression has occurred. The sort of thing that people forget about is leaving boxes in awkward locations, attempting to move heavy office equipment without asking for assistance, opening car doors into traffic, etc. Giving gentle reminders about such things, and offering meaningful tips to avoid trouble, will be worth the effort if just one potential nasty injury can be avoided.

6.16   The Camp Boss

6.16.1   Primary Responsibilities

Similar to the HSE Manager, the Camp Boss has many different fronts and areas of responsibility to deal with, and is reliant upon the inputs from a large workforce spread across a wide variety of job functions to get the actual physical tasks done. The primary responsibilities of the Camp Boss are as follows:

1. Ensure that all utility services are readily available at all times (i.e. water for drinking, cleaning, showering and firefighting, electricity for lighting/power supplies, cable TV, telephone and internet services, etc.), with full backup arrangements provided (such as diesel generators for use in the case of mains electricity power failure, UPS facilities, etc.).

2. Ensure that all fire alarms and firefighting appliances are provided and are fully functioning, and that everything is in place to allow safe evacuation of the camp in the event of an emergency.

3. Ensure that the medical facilities are equipped, maintained and staffed on a 24/7 readiness alert basis.

4. Arrange for necessary catering services to be available at the times needed, and monitor/supervise those responsible for such provision to ensure that the guidelines in respect of health and safety for food preparation, handling and storage are being followed.

5. Arrange cleaning services for the accommodation, cooking, dining and communal areas, as well as for all external areas, and monitor/supervise the workers.

6. Arrange the procurement of and storage space for personal purchases (such as toothpaste, razor blades and other items for personal use of the personnel living in the camp).

7. Ensure that the provision of household consumables such as tissue boxes, soap, shampoos, towels, bed linen, etc. is adequate to support the needs at all times, and that stocks of spare cups, plates, general crockery and utensils are also readily available.

8. Arrange for staffing of all the facilities, including provision of guards if necessary.

9. Arrange vehicles required for general duties such as transportation of food, soft drinks and other products, and ensure that they are regularly serviced and kept in full running order. Where necessary, ensure that adequate vehicle servicing facilities are available, and that waste oil arising from the servicing activities is properly disposed of.

10. Arrange facilities for cleaning, drying and pressing of personal clothing items as well as the general items, such as bed linen, towels, etc.

11. Arrange garbage disposal areas and ensure garbage collection/removal is conducted regularly, as well as arranging for regular septic tanks cleaning and maintenance to be carried out.

12. Ensure that 24/7 maintenance teams are in place to take care of routine and emergency repairs, as well as to maintain the camp facilities.

13. Organise regular drills and training sessions to ensure that all people staying in the camp know what to do in the case of an emergency, such as an outbreak of fire or an armed attack on the camp.

Many of the above responsibilities are in fact routine, day-to-day duties, which should be straightforward to undertake once the initial set-up stage has been completed. Generally, the risks for the Camp Boss are therefore reasonably low, except for the very occasional emergency situations arising, and it will simply require good organisational and monitoring skills to keep everything ticking over smoothly.

6.16.2   Major Risks

The biggest risk for the Camp Boss is with regard to fires that start in the accommodation areas, especially in the night. More often than not, these are related to clandestine food preparation using such things as kerosene burners. This is why it is essential that the fire alarms and firefighting facilities must be regularly checked to ensure that they are in perfect working order. Steps must also be taken to conduct formal training to ensure that the emergency response requirements and evacuation procedures are thoroughly understood by all. Of course, violation of the rules for starting personal fires in the camp buildings must be treated as a very serious offence, and adequate steps should be taken to advise everybody that personal kerosene burners and similar cooking devices are prohibited.

Another big risk can occur with regard to hygiene, where the health of many people can be affected, sometimes disastrously. This can occur if food is not properly stored or prepared, or if one of the kitchen staff has a communicable disease that can be passed on through food. However, there are two things that can be done to reduce this latter risk:

1. Get simple weekly health checks carried out on the entire kitchen staff. Even identifying a high temperature alone can, by itself, help to spot a potential health risk early, without other more invasive checks being carried out.

2. Ensure that polythene or latex disposable gloves are always available to those who will be handling food in the kitchen, and that adequate supplies of fresh, clean water and cleaning cloths too are readily available.

6.17 The Project Information Manager

6.17.1 Importance of Information Management

A colossal amount of information is generated when a major Project is being undertaken. If that information is not organised and stored in a way that enables efficient retrieval and distribution, then it would make it much harder to complete the Project successfully. Not least, this is because the Employer will expect information to be submitted in an organised manner, much of which the Employer's Team will wish to scrutinise and possibly comment on before approval to such information is given. In addition, all the Contractor's Managers will need reliable progress/status information to be made available to them in a timely fashion, so that they can then make the right decisions at the right time. Further, a lot of maintenance and servicing information needs to be handed over formally to the Employer before the facility will be ready to commence operations safely. A good deal of that information will need to be made available even before certain commissioning activities will be allowed to commence (for reasons of ensuring safety of the workers).

6.17.2 Information Overview

Before launching into what the role and responsibilities of the Project Information Manager (PIM) are, I suggest that it is worthwhile looking into exactly what the information is that the PIM is required to manage. That information can broadly be regarded as falling into two distinct categories:

- DOCUMENTS, and

- DATA.

Documents (such as drawings, specifications, technical papers, procedures, letters, etc.) are generally self-explanatory. They do not usually require interpretation in order to understand their content or purpose; they just need to be read properly. Nowadays, the majority of documents used in the construction industry are in electronic form. They are only printed out when needed to be used in situations where use of a computer would be impractical (such as in the field, when construction is being undertaken in line with what is shown on the drawings). Rather than have a dual system of paper documents and electronic documents, paper documents can be scanned and stored in electronic form. The original paper versions can be archived in a secure storage place, just in case they need to be accessed at some point in the future.

Many of the documents produced by the Contractor are what are referred to as 'Engineering Deliverables', and need to be approved by the Employer's Team in a logical, sequential manner. If that approvals system breaks down because of mismanagement on the Contractor's side, serious delays to the Project can occur. It is therefore very important that the Contractor has a means of knowing what document approvals are required, which of those documents have been issued at any given point, and what the approvals status is for each.

Data is of two kinds, required for distinctly different purposes:

1. Project management data, used for the Contractor's management control purposes as well as by the Employer's Team for monitoring purposes. This comprises such information as HSE statistics, work progress figures, etc.

2. Asset/equipment data, used for:

   (i) providing vital information to the Employer's Operations Team, designed to help (a) in the effort to maintain the plant's operational efficiency and availability, and (b) the planned maintenance shutdown work to proceed more efficiently, and

   (ii) the Employer's financial management purposes.

For the Employer's operational purposes, the type of data/information required includes the equipment tag numbers, the equipment characteristics, the spare parts availability and interchangeability details, and the spare parts storage locations, etc. The information required by the Employer's financial team is to aid in such matters as planning/organising the spare parts inventory, and properly accounting for depreciation of the assets in the financial records, etc.

The data that is required for Information Handover purposes (such as the details of the Project's equipment and the corresponding Equipment Tag Numbers) sits inside the Projects' various documents and therefore needs to be extracted. This can sometimes be an arduous and lengthy manual task if the Computer Aided Design (CAD) software does not provide for automating the tagging of equipment and other items. I have observed that manually extracting data from drawings is a task that nobody likes doing, and I have been on a number of Projects where commencement of the manual data compilation and processing began very late. That always ran the risk of delays occurring, due to the added time sometimes necessary to correct data inconsistencies found in the documents (and there will always be plenty wherever manual data collection is required). On a few occasions, commissioning of many critical items of equipment was delayed and, on more than one Project, the completion date was then severely overshot. The need to manually classify equipment and collect tagging data can be overcome nowadays by employing intelligent CAD software; this is a real boon and should be insisted upon by all Contractors from their Engineering Subcontractors.

6.17.3 Electronic Document Management Systems

Today, Employers expect Contractors to invest in appropriate software systems and computer hardware that will handle the Information Management work properly and be fully acceptable to the Employer. Most Employers require that the data produced by the Contractor will be capable of being transferred easily to the software system that the Employer's Operations Team will be using. If that is not the case, then it may require double-handling by the Contractor's Information Management Team to transfer the data already inputted to the Contractor's systems. Alternatively, it may mean that the data must be inputted directly to the Employer's software system. That will almost certainly mean that only a limited number of the Contractor's personnel will be able to access the data that has been inputted to the Employer's server. The Corporate Information Manager should therefore be involved as early as possible in the bidding process and the bid negotiations. In that way, he/she can have a decisive say in how best to set up the Information Management System so that time, effort and money can be saved in handling the Information Management work.

The most basic Electronic Document Management System (EDMS) can overcome the problem of handling documents, by storing electronic copies in a secure location, allowing them to be accessed when needed by authorised personnel. However, the real power of an EDMS comes through if it is one that can be set up to notify persons automatically, both within and external to the Contractor's Project Implementation Team, that a document requiring their attention is ready for their review and input. A good EDMS will manage all electronic information within the defined checking/review workflows (pertinent to what document type is being handled). Not using an EDMS puts a strain not only on the Document Control Team but on the Project Implementation Team as well. On the other hand, using a worthwhile EDMS ensures that the storage and distribution of electronic documents are handled effectively and efficiently, while removing a great deal of the unnecessary strain on personnel.

A worthwhile EDMS will improve transparency of the Project's status, plus enable better collaboration and information sharing between and within the various teams involved. Those teams include the Contractor's Project Implementation Team (especially the leaders in the Engineering, Procurement, Construction and Commissioning Departments), the Contractor's Corporate Management Team and the Employer's Team. The EDMS should enable Managers to obtain standardised reports without effort, and allow the Employer to check easily on the Project's status. To handle the project management and asset/equipment data, many Tier 1 EPC Contractors now employ an enterprise-wide EDMS linked to an Engineering Data Warehouse (in which all the Project's metadata will reside). To manage an electronic information management system that extends across the whole company, a Corporate Information Manager is generally employed. The PIM will report (and provide feedback) at a functional level to the Corporate Information Manager.

It needs to be borne in mind that no 'off-the-shelf' EDMS exists that can be used 'as is' for any company. Each EDMS will be supplied as an 'out-of-the-box' model and will need significant configuration before being useful at even the most basic level; and that takes time to get set up correctly. For those construction companies who have sophisticated clients with special requirements for document and data transfers, major modifications (customisation) may also be necessary to the basic EDMS to bring it to the required stage of refinement. Inevitably, when the EDMS is being applied to a particular Project, there will be many bugs in the system that need to be ironed out. Regular dialogue is therefore necessary between the Corporate Information Manager and the PIM to get such issues resolved as quickly as possible, especially where other Projects are employing similar modifications within the company.

The reality is that setting up an enterprise-wide EDMS is not a cheap undertaking, but the major benefits are that:

  (i) the number of data entry errors are considerably reduced, which eases the strain on the members of the Information Management Team;

 (ii) considerable time can be saved on all future Projects once using it becomes a routine operation within the company; and

(iii) most Employers are very happy to note that such a system is being employed, which gives them confidence that Project data will be handed over in an acceptable form, on time.

6.17.4 PIM Role and Primary Responsibilities

The PIM is often charged with the responsibility of looking after the Document Control Team as well as the Information Technology Team, but the role of the PIM must not be confused with the role of the Document Control Manager (DC Manager) and/or the Information Technology Manager (IT Manager). For the purposes of this book, it has been assumed that the DC Manager and the IT Manager are independent of the PIM, but that there will be the need for close cooperation between all three Departments for which those Managers are responsible.

The PIM's primary responsibilities are as follows:

1. Install a controlled software environment for Information Management purposes, utilising the Contractor's preferred enterprise software systems and tools (if any).

2. Develop an Information Management Strategy that will fully satisfy the Project's information/data collection, storage and retrieval needs.

3. Develop the Information Management Execution and Implementation Plans.

4. Ensure that all the members of the Project Implementation Team understand and support the Information Management Strategy and recognise the savings inherent in effective execution of that strategy.

5. Liaise with the IT Manager and the DC Manager in the development and management of the Project's electronic folder structures and the security access rights across all the platforms used for handling documentation and electronic information/data.

6. Ensure that the Project's naming and numbering systems are fully supported by the Information Management Strategy.

7. Organise/monitor the personnel responsible for inputting the technical data to the various software systems utilised for Information Management data inputting, storage and retrieval, to ensure that the work is being properly handled and completed in good time.

8. Manage the interfaces with the necessary external software support teams.

9. Ensure that Information Handover to the Employer will be able to take place smoothly and in a timely manner, such as will fully satisfy the contractual requirements.

6.17.5 Major Risks for the PIM

The major risk faced by the PIM will be that too many people on the Project Implementation Team will consider that the data management work in respect of the equipment will have been fully completed by the time all the Equipment Tag Numbers (ETNs) have been placed on the drawings, and that there is nothing further to do. However, there is a lot more work to do beyond that, which you might think should be undertaken by the Engineering Team. Nonetheless, there is often a great deal of reluctance within the Engineering Team to accept responsibility for doing that job, even if that reluctance is not voiced outright.

The major tasks the foregoing involves are:

(i) ensuring that the ETNs on the Piping and Instrumentation Diagrams (P&IDs) are faithfully updated to match what is shown on the layout drawings,

(ii) inserting all the ETNs into the management software required by the Employer for asset management and also for operating and maintenance purposes (such as IBM's Maximo or similar),

(iii) ensuring that all Vendor and Subcontractor tagging information is likewise entered into the software management systems, including all 'spare parts' information, and

(iv) making sure that all the associated attributes of the equipment are also faithfully input into the software system.

The big risk here is that the problem of sorely incomplete database information is not discovered until far too late in the day, which then delays the facility's start-up date. This situation can easily occur, because there is always pressure on the Design Engineers to get the drawings and Procurement bidding packages out as fast as possible. On the other hand, dealing with all peripheral data and information recording relating to the ETNs simply slows down the design work progress by taking the Design Engineers away from the pure engineering activities. It is little wonder then that these 'tagging' and 'data/information logging' problems occur to one degree or another on most Projects, and it is exacerbated where a PIM has not been appointed in good time (or at all).

6.17.6 Dedicated Data Inputters

Frankly speaking, there is only one way round the foregoing problem, and that is to provide the PIM with a dedicated, full-time team of people to transfer the relevant data from the drawings and documents into the software management systems. This will have the double benefit of ensuring that the Information Management work will be handled (i) properly and (ii) in a timely fashion. In that way the Engineering Team will be left free to concentrate on producing the drawings, specifications and all other essential documentation required for Procurement and Construction purposes, as well as conduct all the work necessary for the strategies, reviews and reports associated with Technical Bid Evaluations.

As part of their job, therefore, the members of the Information Management Team should be tasked with ensuring that:

(i) the ETNs match the Employer's tagging requirements in all respects,

(ii) the ETNs on the P&IDs match completely with the ETNs on the layout drawings,

(iii) all information regarding associated attributes of the equipment are collected,

(iv) all Spare Parts Interchangeability Report forms have been completed correctly by the Vendors and Subcontractors, and

(v) all the required data and information is properly inputted into the asset management software system as well as the operations and maintenance software system (which may be one and the same system in some cases).

6.17.7 Information Management Support Personnel

A problem for the PIM can arise if the Contractor's proposed team of support Engineers is not sufficiently experienced to be able to do the work required without considerable training from and hand-holding by the PIM. This will be a direct result of the Contractor's Corporate and/or Project Management Teams not fully understanding the functions of the Information Management Team, and thus believing that any Engineer can do the job satisfactorily with minimal training. I have seen many examples of this. The PIM must therefore ensure that the Project Management Team is left in no doubt at all about the experience that is necessary for the Senior Information Management Coordinator role. Such experience must include at least the following, in addition to the selected person having spent a number of years undertaking the role:

1. In-depth understanding and administration experience of multiple Content Management Systems and Electronic Document Management Systems, e.g. Documentum, SharePoint, Wrench, Aconex, etc.

2. In-depth understanding of Engineering Information Management Systems and Enterprise Data Warehouses, e.g. SmartPlant Foundation (SPF), AVEVA NET, Bentley ProjectWise and AssetWise, IBM's Maximo, etc.

3. Experience of Engineering Data Models and Standards, such as POSC Caesar and ISO 15926.

4. Exposure to multiple Engineering Disciplines and Engineering Design tools, e.g. AutoCAD, Intergraph's SmartPlant Designer Tools (such as for Piping & Instrumentation Diagrams (P&IDs), Instrumentation (SPI), Electrical (SPE and 3D)), and Plant Design Management Systems (PDMS) such as AVEVA E3D, Wrench SmartProject, etc.

If the PIM can ensure that all the above systems and personnel are in place in good time, then the PIM's risks will be reduced to the standard day-to-day risks that all Managers must cope with, and which will generally be nothing more than minor nuisances. However, I still see today that some Contractors do not take Information Management seriously enough until they are forced to do so by the Employer's Team.

6.18 The Document Control Manager

6.18.1 Standardised Document Management System

For any Contractor undertaking a major EPC Project, documentation control will be a critical issue, because poor handling of this matter could lead to all sorts of communication problems and even Project delays. Sadly, however, where proper consideration of the possible document control problems usually starts to arise is only after the Contract's signing date. This is when the Document Control Manager (DCM) is brought on board and discovers that the Employer's required system of document handling to be followed by the Contractor is completely separate and different from the system the Contractor had intended using. That situation means that a lot of double handling will be involved in logging all the documents into both systems.

Worse still, those Employer systems and Contractor systems are themselves very often far from comprehensive and may deal only with some particular document types but not others. For example, although drawings and technical documents may be handled well by both systems, such things as tag lists, correspondence, Vendor submissions, Punchlists, etc. may be components of other systems. The right time to have settled this would have been at the bidding stage and during the bid negotiations, when the Contractor should have insisted on one single, comprehensive, standardised document management system (such as, but not limited to, SHAREcat [from Sharecat Solutions AS]). Such a system should also be made readily available to all essential users (although some control will be needed to keep costs down where the system is externally hosted).

Failure to properly manage documentation – i.e. (i) record its existence in a unique but completely identifiable way, (ii) store it so that it can be readily retrieved by authorised personnel, and (iii) automatically distribute an electronic copy to all those in the group for which it was intended – can sometimes result in time-consuming arguments and, ultimately, frustration. Using an effective Electronic Document Management System can significantly improve Project communications by automatically managing all these issues. The more advanced systems allow for documents to be reviewed and commented on, notifying all interested parties as soon as any new documents arrive in the system. They also allow all previous revisions of the documents and the comments thereon (both open and resolved) to be viewed in one place. In addition, they can send out notifications regarding 'respond by' dates and timely warnings if matters have not been actioned/attended to. Not using the power of such systems to the full is a big mistake nowadays, since it opens the door to poor Project communications coming into play. That then leads to the possibility of delays occurring, not to mention possible litigation if the communication problems cause the Project to fail in achieving its primary objectives. On the other hand, if a standardised document management system is employed, the DCM's responsibilities are generally very straightforward with minimal risks attached, provided that he/she is supported by competent staff.

6.18.2 Controlling Communications

It is not the direct responsibility of the DCM to determine who has the authority to issue/sign communications from within the Contractor's organisation. However, it is essential that the whole of the Document Control Department is fully aware of that matter and that nobody makes the mistake of issuing unauthorised communications to third-party organisations or individuals (i.e. all organisations and individuals outside of the Contractor's organisation). Nowadays there are many different ways to communicate with others. It is therefore essential that control is brought to bear on all those working for the Contractor in regard to what can and what cannot be communicated to third parties, and who the authorised communicators are. This same information should also be communicated very clearly to the Employer's Team, with particular emphasis placed on what will constitute valid communications and what must be regarded as invalid communications.

6.18.3 Vetting Documents Before Issuing

For the best level of control, only the Project Manager should be given the authority to issue formal communications to the Employer and the Employer's representatives. Before any letter is presented to the Project Manager for vetting and signing, it is best that the draft communication is sent round formally to all immediately concerned/involved personnel, and finally subjected to what is sometimes termed a 'Squad Review'. The Squad Review of formal communications should be conducted by top-level personnel only, such as the Manager of the Department from which the communication originated, the Senior Discipline Engineer responsible for handling the work that forms the subject of the communication, the Project Contract Manager (or the Contract Administration Manager or the Subcontracts Manager, as appropriate) and the Project Manager. Where the sensitivity of the communication requires it, the Project Director should also be included in the Squad Review. It is even possible that the Contractor's General Manager and legal counsel will be required to advise (but usually only in situations where the subject matter impinges on corporate issues rather than pure Project issues).

The electronic draft version of the communication is the best place to insert a pro forma Squad Review checkbox, which should also have a line for the author/drafter's name, signature and preparation date. Each person reviewing the communication should be required to review the content in hardcopy format, enter his/her respective name legibly, sign and then date their signature to show when the review was conducted. Thereafter, the hardcopy version of the approved draft should be given to the Document Control Department, along with the electronic version of the communication, for the Document Control (DC) personnel to produce the final letter complete with its appropriate unique reference number. When the letter is ready for signature by the Project Manager (or higher authority), the DC member should submit both the Squad Review finalised hardcopy draft and the final unsigned version of the communication to the intended signatory. That person should then sign against the Squad Review checkbox as well as on the final version of the communication.

Adopting such a rigid review method will ensure that no embarrassing communications will ever be issued from within the Contractor's organisation, and it will also ensure accountability inside the organisation for what is issued to third parties. In fact, there is no reason why this method should not be adopted to cover everything that has to be issued by the Contractor (such as drawings, specifications, Materials Requisitions, Purchase Orders, Subcontract documents, procedures, policies, studies, schedules, MOMs, etc.).

6.18.4 Restricting Email Usage

One of the biggest areas of danger for communications nowadays arises with emailing to third parties. The rule should be that nobody below Senior Manager is allowed to send emails to third parties unless the Project Manager has approved the action and, even then, no emails should be allowed to deal with or contain any contractual or commercial issues. Further, all emails should be copied to the DC Department and given a unique formal reference number for recording/filing purposes, and treated with the respect afforded all formal communications. In this way, the Contractor's senior management personnel will be fully aware of everything that is being conveyed to third parties, with no nasty surprises lying in wait for the Project Manager later down the road.

6.18.5 Benefits of a Standard Documents Library

Although it is not the direct responsibility of the DCM to deal with, I have seen companies throw away money unnecessarily by not standardising documentation or by not operating a central server where all documents produced can be accessed for use as templates for future Projects. This should be easy enough to do, and it would save time as well as money. However, I have seen this same problem recurring many times over, where 're-inventing the wheel' was the norm for documents on a new Project. This often resulted in an inferior product being produced that lost the Contractor even more time because the Employer's Team then rejected the submission. The main objection I have heard against standardising documentation is that it would take a lot of time and effort to compile a worthwhile suite of standardised documents, whereas each Project is unique and so those documents would anyway require many modifications.

While I agree with the foregoing view to a certain extent, I feel that the enhanced quality of the end documentation introduced by standardisation would be a far greater benefit for any Contractor in the long-term. My reasoning is that, as improvements are introduced over time, the saving of future time will follow as a consequence. Added to that, the Employers would have greater confidence that the Contractor's submissions are completely fit for the purpose for which they are intended. However, by far the biggest benefit would be that introducing standardised documentation, along with introducing stringent rules about what could be altered, would considerably reduce the risk of compromising content being included in the Contractor's submissions without the Contractor being aware of it; and that can only be a good thing. Where a central library of document templates is in existence in a company, it would of course be the responsibility of the Document Control Team to ensure that the company's templates are being properly utilised, and that unofficial replacement documents that have not been formally approved are not being surreptitiously incorporated into the Project.

6.19 The Information Technology Manager

6.19.1 IT System Security

Of all the management positions in today's construction industry, the Information Technology (IT) Manager probably has the easiest job of all from the point of view of being able to control risks. Do not misunderstand me here, because I am not talking about how specialised or important the IT Team's job is. It is both highly specialised and extremely important, and we could not do our jobs anywhere near as effectively or efficiently without our computerised systems. What I am specifically referring to here is that the primary risk for the IT Manager is the loss of data caused by any means, followed closely by security of the computerised systems from malicious attack. Those risks can be controlled very well, it being just a matter of how much money is available to invest in the necessary hardware, software and backup systems. No two companies will handle IT solutions in the same way, and it will very much depend on the training, overall knowledge and experience of the IT Manager as to what set-up will be recommended to the Contractor's Corporate Management Team. It will also depend on how much money is realistically available for IT purposes.

6.19.2 Backing Up Work from Personal Computers

One of the ironies I have seen about the way IT is handled by many Contractors is that the IT Department is not generally made responsible for ensuring that all work done on an individual's own computer is stored on the Contractors' server. Instead, the work very often resides on the hard drives of company computers loaned (or even 'gifted') to individuals who are scattered throughout all the various Departments of the Contractor. Yet here is a collection of mini-disasters that are just waiting to happen. This is because, if the individuals themselves do not take steps to back up their own data, then all it will take is an unexpected hard drive failure to lose all the work of any one of those individuals. Some companies try to control this situation by operating a centralised server system, which nowadays are veering more towards utilising cloud-based technology for 'Thin Clients'.[29] However, that can prove very expensive to set up and, in any case, such systems are not without their own problems and limitations.

6.19.3 Ensuring Integrity of Computerised Data

Ultimately, the problem for the IT Manager is how to ensure integrity of the Contractor's data in the short, medium and long term. One of the newest approaches to solving the integrity problem at the date of compiling this book is 'hosted computing'. This centralises the processing and storage of data on powerful servers located in a data centre provided by specialist companies. This means that the Contractor would not have to own and maintain the hardware for such an IT system. However, a problem comes

---

29 Ibid., Section 6.9.6.

when personnel have to go to meetings and present computerised information to others, since a laptop or similar computer then needs to be taken along. The integrity issue here is usually dealt with by means of the individual having to log into the company's server remotely, but this relies on there being a satisfactory Internet connection in that location (particularly in regard to bandwidth availability). This is often not possible in some overseas locations, and thus the individual must rely upon using a standalone laptop, downloading the necessary data from the company's server and then uploading any additional/new data after returning to base. The Contractor's server should therefore be equipped with the strongest possible malware checking software to prevent viruses, worms, Trojan horses or spyware being uploaded from the computers of individuals, which otherwise will probably not have been provided with security software to the strength level required nowadays.

To deal with the foregoing risks effectively, it will also require procedures to be implemented to ensure that vital information/data is not lost to the Contractor through any cause, including the departure of personnel from the Contractor's employment. One way round this is to insist that the IT Department should be entitled to go to any individual whose computer is used for the Contractor's business purposes and download a copy of all data files other than personal files. This activity could be set for a given date each month for each individual. This step may sound controversial at first. However, if all employees are made fully aware of the reasoning behind the action and can be forewarned to remove their personal data if it is sensitive information, then there is no reason why this important step to ensure the integrity of computerised work data should not become the norm. Where the Contractor has 'gifted' computers to all its staff, this requirement should be less controversial.

6.19.4 Reducing Contractor's Software Costs

A big problem nowadays is that software piracy is, quite correctly, being clamped down on across the globe, and the last thing that a Contractor needs is to be subjected to a raid from third-party compliance personnel that results in the Contractor's servers and/or individual computers being seized. However, providing up-to-date software for all the administrative and engineering workforce would not only be very expensive, but it would also be a logistical nightmare for the Contractor's IT personnel to keep on top of.

This is why the appointment terms of some companies stipulate that its staff will be provided with a brand-new, free computer upon starting work (and which will be renewed every, say, three years). That computer then becomes the staff member's personal property. However, that then requires the staff member to pay for and load the software that will be needed for doing their daily work onto that computer. Further, all maintenance and repairs for the individual computers become the sole responsibility of each respective staff member. This arrangement has the advantage for the Contractor that it would cut down on the bill for providing software for all the individual computers. More importantly, it would also cut out the need for the larger IT workforce that would otherwise be necessary to maintain the functionality and well-being of all those computers. The disadvantage, as mentioned earlier, is that steps need to be taken to ensure that all essential files are uploaded regularly from the computers of the staff members to the Contractor's central server, especially when employees leave to work elsewhere.

6.20 The Interface Manager

6.20.1 Primary Function and Responsibilities

Interfaces are where two different sections of work undertaken by two different sets of people need to be connected together. Very often those sections of work are being undertaken by two different Subcontractors/Vendors, and the task of organising those interconnections does not happen by itself. More usually, each Subcontractor/Vendor will tend to focus primarily on the work they are engaged to do, leaving the coordination work to the Contractor. That coordination usually takes the form of raising information requests or obtaining clarifications/documents from another party, where the responses need to be available by a given critical date or otherwise risk delaying certain key elements of work.

Without that coordination work being arranged in good time, the Project itself may suffer serious delay. This is perhaps why such matters are referred to as 'Interface Issues', and where the Interface Manager comes into play. However, except on a huge Project, the Interface Manager will most likely not be appointed to do that coordination work on a fulltime basis, simply because the volume of work involved may not warrant it. Nonetheless, the role of the Interface Manager is vital in order to ensure that all utilities will be hooked up properly by the required time.

The Information Manager's first task is to develop an Interface Management Plan designed to avoid or minimise possible conflicts, disruption and/or delay arising from Interface Issues, and to eliminate gaps or missing elements at the interfaces during development of the Project's Engineering Deliverables.

6.20.2 Interface Management Plan

The Interface Management Plan must complement the construction plan, and it must be updated from time to time to reflect changes in requirements or to incorporate improvements. The key objectives of the plan are to put procedures in place that will:

1. Clearly define all interface management roles and responsibilities, and ensure that they are understood by all the Project's personnel.

2. Ensure that all Interface Issues are identified as early as possible and are properly documented and managed through a structured process, including updating of outstanding Interface Issues on a regular basis (through use of an Interface Management Register).

3. Ensure that communications are cooperative, proactive, clear, accurate, timely, consistent, and focused on resolving the Interface Issues.

4. Ensure that Interface Issues are resolved quickly in support of the Project's objectives.

5. Quickly identify those Interface Issues that are likely to negatively impact the Project's costs or the Project Schedule, and allow early communication of the effect of those impacts to all stakeholders.

6. Clearly define and obtain agreement to all interface requirements, including the Interface Deliverables and due dates.

7. Allow easy identification of non-performing Interface Issues, and provide for their escalation to a higher level of management that may be required for resolution.

8. Provide for verification, measurement and continual improvement of the effectiveness of the interface management process.

The key to successful management of Interface Issues on a Project will be to establish well-defined and properly structured procedures for managing those interfaces. The Interface Management Plan must therefore clearly specify the methods for identifying Interface Issues, defining the scope and the Deliverables for the interface activities, assigning responsibilities for managing the Interface Issues, monitoring the status, and following-up and closing-out of those Interface Issues. It should be borne in mind that resolving an Interface Issue during the Detailed Design phase will be far more cost effective than leaving it until the Construction phase.

The role of the Employer's Team ought to be to review/monitor the Contractor's interface activities and, if it becomes necessary, assist the Contractor by facilitating a resolution of non-performing Interface Issues. However, it is sometimes very hard to get the Employer's Team to respond appropriately, since the general thinking from the Employer's side seems to be that all such problems are for the Contractor to resolve. I have observed that this tends to be more of a problem if a Project Management Consultant is involved than if the Employer's own in-house team is managing the Project.

6.20.3 External and Internal Interfaces Explained

Interfaces between different groups working on a Project can be classified as either 'external' or 'internal' interfaces, but I have observed that, for some people, this classification is not so easy to determine. The external interfaces are the prime concern of the Interface Manager and comprise third parties that may or may not be operating under Purchase Orders or Subcontracts instituted by the Contractor. The internal interfaces will generally occur between the different Departments involved on the Project, and they must be managed by the Managers of those Departments, not by the Interface Manager.

In that regard, it is important to understand that any party contracted to provide services, undertake work for or supply/deliver materials, goods and/or equipment to the Contractor must generally be regarded as an internal party, not an external party. For example, the consultant company undertaking Engineering services for the Contractor must be regarded as an internal party, and thus the responsibility for managing the interfaces between the Contractor and that entity falls to the Contractor's Engineering Manager, not the Interface Manager. Likewise, management of the Contractor's administrative interfaces with the Vendors and Subcontractors appointed for the Project are, respectively, the responsibility of the Procurement Manager and the Project Contract

Manager (or the Contract Administration Manager, Subcontracts Manager and similar such persons).

However, there is one very important exception to the above general rules, and that relates to where interfaces occur between Vendors and Subcontractors themselves in regard to the physical work interfaces, which should be considered to be external in nature. This is because the actual on-Site work for such interfaces will most likely occur after the main Engineering and Procurement activities have been completed. That means that there will be nobody from the Engineering and Procurement Departments to follow up any outstanding issues for that work. The Interface Manager will be required to follow up such interfaces, and also manage the Interface Issues that occur between the Vendors and Subcontractors with the Construction and Commissioning Teams. This requires the Interface Manager to remain in position until handover of the completed facility has been achieved.

6.20.4   Identifying/Locating Interfaces

For those who have never been directly involved in dealing with the entirety of a Project's Interface Issues, I should perhaps explain here that the majority of the interfaces for the physical (on-Site) work can generally be identified from the Piping and Instrumentation Drawings (P&IDs). This is because the P&IDs usually show where the battery limits of the various work packages are, along with all the resultant tie-in points. Those interfaces will generally be with and between Vendor packages or the Contractor's or Subcontractors' work, and also, to a lesser extent (but by no means less important), with external utilities suppliers, such as electricity and water authorities.

From the information shown on the P&IDs, the Process Engineering Team should be requested to compile a list of the tie-in points (preferably in electronic format). That list should then be passed over to the Engineering Manager for transmittal to and discussion with the Interface Manager. The Interface Manager would then be able to utilise that data to compile the Interface Management Register (IMR). The IMR is a matrix identifying each individual Interface Issue (very often in spreadsheet format), where the key information for the purpose of tracking the processing and closing out of the Interface Issues is logged.

From time-to-time, as changes to the P&IDs occur, more Interface Issues may arise, or the nature of the existing known Interface Issues may change. If that occurs, then it will be the responsibility of the Engineer involved to notify both the Engineering Manager and the Interface Manager as soon as possible regarding all such changes that are uncovered, and the IMR should then be updated accordingly. However, it must be remembered that any spreadsheet employed must be made available on the Project's dedicated computer server so that ownership can be shared. Leaving ownership of the IMR in the hands of just one person can lead to all sorts of problems. This is because, as mentioned earlier, updating it is not a fulltime job and the Interface Manager may well have another important role to play on the Project. However, the IMR does require being followed up on a regular basis with all others responsible for inputs, in order not to let outstanding essential inputs drift on for too long. Others should therefore be made available to assist the Interface Manager to do the necessary chasing and updating if he/she is engaged on other pressing activities.

6.20.5   Interface Slip-Ups

The biggest mistake I saw made with regard to Interface Management was for the Project Manager not to take the work too seriously and thus allocate somebody to perform the interface coordination task as a temporary job only. Later, that person was then moved to another job, since the Project Manager had forgotten that the staff member concerned was also dealing with Interface Management 'on the side'. On one Project I was involved with, the temporary Interface Manager retired, and all the data was wiped clean from his desktop computer, on which the sole version of the spreadsheet had been sitting, with no hard copy having been made available. Playing catch-up under those circumstances proved very hard to do, and was accompanied by a severe loss of face for the Contractor.

On another occasion, the provision of eight pumps falling between the battery limits of two specialist equipment packages (and which could have been supplied by either of the two Vendors involved) had been forgotten about as an Interface Issue by the Engineering Department. This was despite the fact that the interface point had been clearly marked on both sets of Vendor drawings. Unknown to the Construction Team until almost too late, neither of the two Vendors had been requested to supply those pumps due to a mistake on the separate sets of drawings supplied to each of the Vendors. They had in each case shown the pumps sitting outside the battery limits of both sets of the separate Vendor equipment. Those pumps were specialist in nature and required six months manufacturing time, added to which the factories in Italy were about to shut down for their one month's annual holiday. With just three months to go before the commissioning of both sets of equipment was due to commence, it could have been a disaster for a smaller Contractor with no 'buying power' to assert. Luckily, the Contractor I was involved with had sufficient pulling power to overcome the delay problem, although not the monetary problem. However, what had caused the problem in the first place was that the Interface Manager had been working primarily in another position with the Commissioning Team, and had not had time to track the Interface requirements to spot that the manufacturing/ordering of the pumps had not been attended to.

An even bigger surprise came with a hospital Project I was involved on, where an incinerator had been planned for in order to dispose of the toxic waste arising from the hospital's activities, the design for which was duly incorporated into the drawings. It was only late in the day that it was found that the Local Authority would not countenance an incinerator for toxic waste in that location (an outstandingly high-quality residential area). That meant that the hospital had to arrange for separate storage for all toxic waste and remote disposal in specially designed facilities, using suitably experienced transportation companies with appropriately equipped vehicles. The problem was that nobody had been appointed as the Interface Manager until much later than should have been the case, and speaking to the Local Authority in the early stages of the Project had somehow been overlooked. However, a similar situation could arise for any type of plant that produces toxic waste in the pharmaceutical or petrochemical businesses.

The moral of the above stories is that Interface Management is a serious subject that must not be forgotten about, and which needs to be considered right from day one. This is the only way to ensure that any unexpected surprises are discovered in time for remedial action to be taken to safeguard the Project's required completion date. One way to keep Interface Management in the forefront of everybody's minds that I have witnessed is to insist that it is included on the agenda for every weekly progress meeting. A hard copy of the Interface Manager's updated spreadsheet should be tabled at each meeting and attached to the meeting minutes. In this way, even if the Interface Manager is doubling up on duties and is then moved on elsewhere, there will be a permanent record of the current status available for the next person in line to take on the responsibilities of the Interface Manager.

6.20.6   Closing Interface Activities is Hard Work

External interfaces on a Project will also occur in respect of utility service providers, such as electricity, gas and water services suppliers. In theory, those service providers would be expected to be eager to supply the required services and thereby earn additional income for their companies. However, in overseas locations where those entities are government-owned corporations, it is much more likely that cooperation in getting the required services will take far longer than anticipated. Such problems usually arise due to a combination of such things as lack of finances for capital expenditure within those entities, plus the lack of skilled manpower. Sometimes, political in-fighting amongst the various government agencies involved is the cause. If the Contractor's connections to such utilities require temporary shutdown of the services to other users, this makes the matter even more complicated. On one of the Projects I worked on, the 'urgently required' Overhead Transmission Line (extending a great distance) had been installed and tested but had to wait for 18 months to be connected to the electricity service provider's substation. This was due to there having been no alternative supply available to the existing consumers while the substation was going to be out of action for two full weeks while the hook-up activities took place.

Even with a supposedly simple matter such as hydrotesting of pipelines, actually getting the water supplied in the quantities required can be very difficult and take much longer than expected. If pipe preservation precautions are then required (usually by using additives in the test water, owing to the long time it will take to fill the pipeline), the next difficulty will be getting permission to discharge the water. This can also take a very long time while tests are conducted to ensure that the discharge will not cause environmental damage. The list of such problems is endless, and overlooking them can be a disaster for the Project when those problems come very late in the day.

The only way to deal with these situations is to start a dialogue with the service providers and concerned Local Authorities as early as possible, in order to find out what needs to be done to get the necessary connections (or permissions) in place by the required date. Leaving such Interface Issues to the last moment will almost certainly result in such connections or permissions being made much later than required. In addition, there will be no valid excuse for the Contractor to claim extension of time with the expectation of the associated additional costs being paid for by the Employer.

6.21 The Contract Administration Team

6.21.1 Determining the Compilation of the Team

On a large and/or complex Project, the Project Manager would usually be supported by a full Contract Administration Team (CAT). That team's function would be to monitor and administer the contractual and commercial aspects of the Project. It would also entail advising the Project Manager as to how best to protect the Contractor's contractual/commercial interests, as well as dealing with any associated problems that arise. On the other hand, on a very small or straightforward Project, the Project Manager alone may well undertake the contract administration work, without any support team being necessary. If the Project Manager does take on the contract administration role, then it is vital that the responsibilities, risks and pitfalls of doing so are fully understood. The remainder of Section 6.21 explains this in more detail, although not all issues discussed may be required on a very small Project. Having said that, some Project Managers may be surprised to see the extent of responsibilities covered by a full CAT and decide that taking on that role is a bridge too far for them.

Where a mega Project is involved, the CAT may comprise a number of Contract Administration Managers and, possibly, Subcontracts Managers too, each of whom would be responsible for monitoring and administering a different segment of the work. Examples of such different work segments would be elements of work in locations remote from the main Site area, such as off-shore structures and facilities, intermediate compressor or gas/oil pumping stations, block valve stations on a long pipeline, large pipe racks running over long distances, large-scale specialist processing equipment, overhead transmission lines, power generation facilities, SCADA (Supervisory Control and Data Acquisition) installations, etc.

Further, if the size and complexity of a Project warrants it (which would not normally be the case for anything under USD 2 billion), a Project Contract Manager may be appointed to head up the CAT (as distinct from the person heading up the corporate Contracts Department). The actual role and responsibilities of the Project Contract Manager may vary, dependent on the nature of the Project and what (if any) Contract Administration Managers and/or Subcontracts Managers are appointed. No two Contractors seem to follow the same approach to contract administration. However, as long as the split of roles and responsibilities is well understood, as well as to whom the responsibilities fall for managing the risks involved, it does not matter too much how this split is made or what the position is called.

To a large extent, therefore, it can be appreciated that the roles of the Project Contract Manager, the Contract Administration Manager and the Subcontracts Manager overlap. Consequently, the risks faced by a Project Contract Manager under the all-in-one role can be considered to be an amalgamation of those described within this section (Section 6.21). The Contractor must therefore give due consideration at the bidding stage to the size and complexities of the Project in order to determine the specific positions to be allocated for the personnel comprising the CAT, as well as their respective roles.

6.21.2 The Project Contract Manager

The primary responsibilities of the Project Contract Manager would generally be as follows (assuming that Contract Administration and Subcontract Managers are also appointed):

1. Organise the work of the Contract Administration Department and direct and monitor the activities of the Contract Administration and Subcontracts Managers to ensure that they remain in line with the Contractor's policies and procedures.

2. Oversee the preparation of the Bid Enquiry Documents and Subcontract documents for all potential Subcontracts.

3. Assist the Procurement Manager in preparing the standard Request for Quotations documentation to be issued to potential Vendors, including perhaps drafting (or at least vetting) the 'Terms and Conditions of Purchase' to ensure that all flow-down clauses required by the Employer will be properly incorporated into the final Purchase Orders. That same exercise would need to be done in respect of the contractual documentation for the Subcontractors.

4. Advise the Project Manager, the Project Controls Manager and the Procurement Manager regarding potential contractual Claims or Disputes between the Contractor and any Vendors and/or Subcontractors as soon as such issues become evident, as well as how to handle such matters in the most effective manner.

5. Assist the Contract Administration Managers and Subcontracts Managers in their efforts to resolve all Claims and Disputes effectively and efficiently, based on the objectives decided by the Project Manager.

6. Review all incoming correspondence from the Employer's Team to check for problems, such as additions to the Project Scope or complaints of any kind that need to be formally addressed and resolved.

7. Vet all outgoing correspondence from the Contractor's side, in order that the language used is appropriate, not offensive and does not compromise the Contractor's contractual or commercial position.

8. Be the spokesperson for the Contractor in regard to all contractual and commercial negotiations with the Employer's Team.

9. Write up the originals (or finalise the 'position papers' prepared by others) for legal review when it becomes apparent that a Claim or Dispute may well be heading for litigation/arbitration.

10. Be the focal point for liaison with the Contractor's Legal, Finance and Insurance Departments on all legal, financial and insurance matters (e.g. Performance Bond, Parent Company Guarantee, Letters of Credit, Retention Bond, insurance claims, etc.).

6.21.3 The Contract Administration Manager

Where a Contract Administration Manager (CAM) is appointed, the following would generally be his/her primary responsibilities (assuming that a Project Contract Manager and at least one Subcontract Manager have been appointed):

1. Compile a list of the Project's Deliverables, and monitor the documentation preparation to ensure that each Deliverable is submitted to the Employer in a timely manner.

2. Compile a Contract Summary of all the key obligations of the Contractor, and disseminate it to all other Managers on the Project.

3. Ensure that all Managers are fully aware of their contractual obligations and how to discharge them properly, ideally through the medium of conducting meetings to explain what is set out in the Contract Summary.

4. Ensure that all Managers are fully aware of the Project Scope and the need to keep both the Project Manager and the Contract Administration Manager aware of all Claims and Disputes as soon as they arise.

5. Manage the Subcontracts in conjunction with the Subcontracts Manager(s) and the Cost Management Team, including handling Variations and Claims issues.

6. Manage Variations arising from the Employer's instructions or unexpected Site conditions.

7. Keep the Project Contract Manager and the Project Manager informed on all contractual Claims and Disputes arising between the Contractor and any Subcontractors, and assist in compiling all the information/data necessary in connection therewith, in order to be able to defend the Contractor's position or fight the Contractor's cause successfully.

8. Draft up all documents to support Claims and Disputes for the Project Contract Manager to review.

9. Draft up final 'position papers' for review by the Project Contract Manager in respect of all Claims or Disputes that are headed for litigation/arbitration, or assist others to do so where that may be the more appropriate action to take. This latter approach may be preferable where the arguments hinge on extremely complex technical details that require inputs from specialists.

6.21.4 The Subcontracts Manager

Where many Subcontractors are involved on a Project, a number of Site-based Subcontracts Managers may be appointed. The primary responsibility of the Subcontracts Manager is to handle all the administrative work for the subcontracted work packages, including the measurement work involved in the evaluation of Variations. In addition, the Subcontracts Manager is required to measure and record the monthly progress of the Subcontractors for interim payment purposes. Very often too, the Site-based Quantity Surveyors and/or Cost Engineers will then be required to operate under the control of the Subcontracts Manager(s) rather than the Project Controls Manager or the Contract Administration Manager. Such cost-orientated specialists will usually be the ones responsible for undertaking the detailed measurement and pricing work for the Variations.

As mentioned above, the roles of the Contract Administration Manager and the Subcontracts Manager tend to overlap to a certain extent and, where they do, some of the responsibilities and risks mentioned above as being the Contract Administration Manager's responsibility may then fall upon the shoulders of the Subcontracts Manager to deal with. The exact split of responsibilities should be decided by the Project Contract Manager (if one is appointed) or, otherwise, the Project Manager.

6.21.5 Compiling the Contract Summary

Whether the Project Contract Manager or the Contract Administration Manager compiles the Contract Summary, its contents must include appropriate details for at least the following topics, so that every Manager may have a 'quick reference' document by his/her side explaining all the major elements of the Project and the Contractor's obligations:

(i) Phases of the Work,

(ii) Work Packages,

(iii) Contract Documents,

(iv) Completion Times/Dates,

(v) Transmitting Project Information to Company,

(vi) Early Project Deliverables (with a list of those needed within the first 30/60 days),

(vii) Meetings and Progress Reporting Requirements,

(viii) Plant Performance Testing,

(ix) Liquidated Damages,

(x) Guarantee/Bond Requirements,

(xi) Insurance Requirements and Liabilities,

(xii) Warranty Period,

(xiii) Final Acceptance Certificate, and

(xiv) Procedure for Handling Change Management and Potential Variations.

It is important that the Contract Summary is written in such a way that all the Managers will readily appreciate its usefulness and thus be inclined to keep it by their side for a while at the commencement of the Project (at least until they have absorbed all the key points).

6.21.6 Identifying Risk Exposure

Amongst the issues that the Contract Summary should specifically identify and draw the Project Manager's and Departmental Managers' attention to is the extent of the risk exposure that the Contractor may have in respect to:

(i) any areas to be developed in the Scope of Work;

(ii) the applicability dates of the specified codes and standards (especially where they are not limited to the date of Contract signing);

(iii) the priority of the documentation in the event of conflicts between the contents/requirements (as will most likely be specifically mentioned in the Contract);

(iv) any conflicts that have been found in the Contract Documents;

(v) whether or not the Contract supersedes all other prior agreements (if any);

(vi) whether or not the Project Schedule is tied to Liquidated Damages;

(vii) whether or not Liquidated Damages are the sole remedy for being late and are limited in the aggregate (or whether termination is an option for the Employer if all Liquidated Damages become payable);

(viii) whether Liquidated Damages are due for things other than delayed completion, such as failure to satisfy performance guarantees;

(ix) whether or not the Employer's approval is needed before placing Subcontracts and/or Purchase Orders;

(x) which party bears the risks of inclement weather;

(xi) whether or not unverified/unverifiable Site conditions have been accepted by the Contractor;

(xii) whether or not the Employer requires specific 'Hold Points' in design and/or construction work;

(xiii) whether or not the Contractor has liability for the performance of others (third parties) on the Site;

(xiv) the impact of delays on the Project Schedule caused by others over whom the Contractor has no control;

(xv) the adequacy of any 'Force Majeure' provisions in the Contract; and

(xvi) with whom the responsibility for obtaining licences and permits rests, not just for the Project itself but also in respect of immigration visas and work permits for the workers, special permits for blasting and any other dangerous activities, etc.

6.21.7 Importance of Keeping Records

The primary risk faced by the person/team responsible for heading up the handling of the contract administration work is that, in the event of a potential Claim scenario, others in the Contractor's Team may not have kept adequate records to support the Contractor's position, thus making it harder to win a contractual battle. The impacts of that risk occurring can be reduced significantly by the appropriate CAT member following up closely on all potential contractual and commercial problem areas without loss of time, and chasing up those others for any outstanding information, data and records while the topic is still current. If this is not done as a priority, then leaving it until later may well mean that a lot of valuable information and data are permanently lost. That can easily happen if somebody with all the knowledge leaves the Project before his/her inputs have been obtained.

If a conflict does require litigation/arbitration to be brought into play in order to resolve a particular matter, then the Contract Administration Manager (Lead) or the Project Contract Manager (if such appointment is made) would usually be the person given the responsibility for compiling, into 'position paper' format, all the information/data necessary to defend the Contractor's position or fight the Contractor's cause successfully. He/she would also be expected to substantiate the position/case in front of the lawyers responsible for appearing before the judge or arbitrator. As mentioned above, this can often be where the biggest risk for the Contract Administration Team's personnel sits, in that the required information/data is sadly lacking when it is called for. To avoid this embarrassing situation, the Contract Administration Manager or Subcontracts Manager (as appropriate) must chase up hard for all relevant information that can be made available to be tabled as quickly as possible. A dossier on each and every potential Claim or Dispute must also be compiled as soon as it becomes known about, and it must be added to as and when any new information/data surfaces, so that a sound case for a Claim can be constructed without delay.

A register of all such potential Claims and Disputes must be maintained and updated regularly to show the current status on a real-time basis. Ideally, electronic copies of all relevant material should be made by the Contract Administration Manager and deposited in a protected folder on the Project's server, which the Project Contract Manager and Project Manager must be able to access at all times. It is preferable for such data to reside under individual folders for each head of Claim or Dispute.

Personally, I also find it very useful to keep hardcopy files with the same background information in, simply for ease of reference. However, in this age of tree conservation in particular, I admit to feeling a little guilty about it, despite the fact that I will hand over those files at the end of the Project for archiving purposes. The reason I still insist on maintaining hardcopy files is that they allow me to keep separate chronological records of each individual topic, along with all relevant notes and spreadsheet calculations compiled both by myself and others, together with both 'in and out' correspondence, MOMs, quotations, invoices, etc. This enables me to read through the individual files as full and complete 'storylines', which simply is not possible from the individual Project files held on the company's server. This approach has the very real advantage that I can see very quickly, almost at a glance, exactly what the latest state of play is for any topic. Added to that, those files are extremely useful, and save a lot of otherwise hard work, when it comes to the preparation of 'position papers'.

An added bonus of maintaining hardcopy files is that it also gives me somewhere to store all those paper copies of documents that we all still seem to be so fond of. That then enables me to maintain a 'clean desk' policy at all times, thereby avoiding that mountain of paper that clutters up the desks of some of my colleagues. This is important to me, because I am a firm believer that a clean desk is the sign of an organised mind. This is despite the fact that I still do have reluctant admiration for a few of my former colleagues who knew exactly where to find any piece of paper they needed, despite the seeming clutter on their desks.

6.21.8   Risk of Collusion in Measurement/Costing Team

The major risk for the Subcontracts Manager is that there may be collusion of a fraudulent nature between the Subcontractors and the Contractor's Quantity Surveyors and/or Cost Engineers. This may involve situations such as the over-valuing of work-in-progress (usually in order to achieve a better cash flow for the Subcontractor) or the over-pricing of Variations work (by falsely increasing the quantities and/or manipulation of the prices). It is therefore essential that checks and balances are built into the evaluation process to reduce the chances of this situation occurring.

The best way to do that is to have another Manager outside of the Contracts Administration Team conducting and signing off against the validity checking of the outputs from the Variations assessment and progress evaluation processes. Due to the possible conflict of interest that may occur, it is best not to appoint the Project Contract Manager, the Contract Administration Manager, the Subcontracts Manager, the Quantity Surveyors or the Cost Engineers (or anybody in their respective sub-teams) to conduct the validity checking. The best people for undertaking that task are, I suggest, members of the Project Controls Team, who do not generally have any direct contact with the Subcontractors' personnel. This will quite obviously increase the time necessary to complete the full payment process, but it will have the benefit of ensuring that all steps in the assessment and evaluation processes will be checked more meticulously. This should then not only enable any anomalies to be spotted much easier but also reduce the opportunity for shenanigans to occur.

6.21.9   Facilitation Role of the Contract Administration Team

A well-known international Contractor in the Oil and Gas Industry that I was associated with was mainly headed up by lawyers, and its Corporate Management Team did everything in its power to ensure that the Contracts they signed up to were set up so that all parties were fully aware as to what needed to be done, by whom and by when. The Corporate Management Team of that company made it clear to its Contract Administration Teams that its philosophy was to find win-win solutions to contractual problems, and to avoid escalating issues to the level of litigation/arbitration if it was reasonably possible to do so. In short, they expected the Contract Administration Managers to be facilitators, not trouble-makers.

If any particular issue looked set to turn ugly, the Contract Administration Manager was expected to keep the Corporate Management Team fully informed via the Corporate Contract Manager (or Project Contract Manager, if one had been appointed). Thereafter, the Corporate Management Team would take over the reins and handle the matter in conjunction with the Project Manager, supported by the Corporate Contract Manager and the Contract Administration Manager and/or Project Contract Manager. Unnecessary personality clashes are sometimes the cause of many Disputes turning uglier than they should. However, that particular company provided one of the most successful set-ups I have witnessed for resolving conflicts without personality clashes being allowed to come into play too much at the Project level, largely because of the 'facilitator' role played by the Contract Administration Managers.

6.21.10   Illusory Claims

One problem the Contract Administration Team may face is that the Project Manager or higher ranking person insists that a Claim exists where one truly does not. Telling the boss he/she is wrong can be worse in some cultures than others, but it is never easy under any circumstances. The risk a Contract Administration Manager (CAM) faces by telling the truth directly is losing his/her own job or ruining his/her chances for promotion. However, in some more enlightened major contracting entities, the Project Manager is himself unable to remove the CAM directly, since the CAM is often perceived by the Board of Directors as fulfilling a very necessary 'checks and balances' function.

The best way around avoiding a confrontation with the more senior person would be for the CAM to take the problem away to think about. Having then considered the matter in depth, he/she should set down all the pros and cons of the proposed Claim in writing. This should be approached in the same way as when presenting a 'position paper' to an arbitrator (i.e. using non-emotive, balanced arguments), in order to demonstrate categorically why the proposed Claim could not succeed. If the more senior person still insists on going forward with what the CAM has demonstrated is an unwinnable case, then at least the CAM will have protected himself/herself from criticism in the future. Such protection (which is actually just a 'cover your rear-end' approach) will be very valuable if pursuing the Claim goes badly wrong and costs the Contractor further money that is subsequently found to be unrecoverable. I have been thankful on quite a few occasions that I took that step.

6.22   The Human Resources Manager

6.22.1   Primary Responsibilities

The following sets out the primary responsibilities of the corporate Human Resources (HR) Manager in regard to Project work (not the corporate role), which must be conducted in conjunction with the personnel working in the corporate HR Department in accordance with the HR policies, procedures and processes set by the Contractor's Corporate Management Team:

1. Ascertain the Project's manpower resourcing requirements from the Project Manager for all administration and supervisory personnel as well as the labour workforce (after the Project Director has approved the numbers), together with the start dates required, and establish the level of expertise and qualifications required from potential candidates.

2. Forward the manpower resourcing requests (as approved by the Project Manager) to the appropriate personnel in the HR Department, and follow up to ensure that an appropriate appointment for each position has been made within the required time-frame.

3. Recruit new personnel through the approved channels and ensure that the employment agreements are satisfactory. In regard to this, the HR Manager must keep in mind that the choice as to which candidate to appoint is the prerogative of the Project Manager in conjunction with the Project Director, and not the HR Manager's decision.

4. Establish and maintain personnel files and records.

5. Make arrangements for and get approval of the travel and accommodation requirements of the personnel (both existing staff and newly recruited personnel).

6. Arrange for orientation of new employees, both at the home base and in the overseas country in which the Project is located.

7. Ensure that employee time logging, overtime booking and all other essential record keeping for salary payment purposes is properly set up to allow prompt payment of salaries by the due dates.

8. Calculate the salaries due, and ensure that salaries are paid on time.

9. Organise the holiday roster and approve employee leave arrangements, all in discussion with the relevant Department Managers.

10. Maintain vacation and sickness leave records, etc.

11. Monitor the performance of employees for the purpose of internal appraisal, promotions, remuneration adjustments, etc.

12. Assess the needs of essential skills training for employees, and arrange for such training as necessary.

13. Arrange the transference of personnel to other jobs in a timely manner when their assignments are coming to an end.

14. Deal sympathetically with employee issues as they may arise from time-to-time, and discuss/advise for resolution with the relevant Department Head and the Project Manager, in line with the employee relationship plans.

15. Manage the insurance arrangements for employees, including Employers' Liability Insurance, Medical Insurance, etc.

16. Handle employee relations and deal with sensitive administrative issues, including matters such as sickness, absences, complaints/grievances and disciplinary issues.

6.22.2 Becoming the Employer of Choice

Assuming always that the remuneration levels applicable to each role are well understood and actually paid by the Contractor, the biggest risk faced by the HR Manager is when it becomes an employee's market, not an employer's market, and finding suitable candidates within the required time-frame becomes very hard to do. If this coincides with a sudden upswing in the Contractor's workload, failure to place people in position on time could then cause serious problems for the Project. An upturn in the market will inevitably mean that the Contractor will be competing with everybody else out there to secure suitable new employees. It is therefore very important that the Contractor should have worked hard in advance to cultivate the image of being a great employer, so that it stands out from the crowd. However, what a lot of Contractors fail to realise is how things have rapidly changed over the past decade, and that people now happily share their good and bad work experiences online with their friends.

The foregoing means that, if the Contractor that the HR Manager works for has not been careful with its handling of employees in the recent past, then there could be a lot of negative comments about the Contractor out there on social media and job-seeker platforms. If so, that will make recruiting the right people even harder. It is therefore the HR Manager's job to continuously steer the Contractor's Corporate Management Team towards fair treatment of its employees at all times. This means that the Contractor must at all costs avoid acting in such a way that it is hauled before Labour Courts/Tribunals to explain its apparent unfair actions. It does not matter how right the Contractor was; the fact that many ex-workers take the Contractor to task publicly is not good for the Contractor's future recruitment drives. This is regardless of the fact that you may have heard people say, for example, 'There is no such thing as bad publicity'.

The one important matter that the HR Manager therefore needs to instil in the thinking of the Contractor's Management Team is that the HR function does not end in recruitment. Far from it, because a lot of time, energy and money goes into teaching employees all about the workings and intricacies of an organisation. It is therefore essential that great effort is put into retaining good employees. This is because they may find it all too easy to leave an organisation if no real attempt has been made to foster good relations with them and to engender a genuine sense of loyalty and belonging inside the company. Steps in the right direction include offering first-class training programmes and career progression opportunities, etc. It also helps to be seen seriously considering the welfare of the employees, such as watching out for and helping stressed-out workers, etc. Mouthing the words 'our workers are our strength and best asset', but then doing nothing to show belief in that assertion, does not go unnoticed, both within and outside an organisation.

6.22.3   Recruitment Issues

I have seen Contractors attempt to handle recruitment on their own, usually through setting up a website showing job openings, and also by advertising in local newspapers. The website approach relies upon potential candidates knowing that the advertisement is there or stumbling across it accidentally, and is not a sound means of recruiting in great numbers. On the other hand, advertising in newspapers can sometimes bring in thousands of responses, but where the bulk of them will inevitably be 'hopefuls', offering nothing of real value and bogging the recruiting team down in worthless additional paperwork. Such a situation can lead to a lot of unnecessary stress and late night working for the staff in the HR Department, especially where the candidates are in overseas locations and in vastly differing time zones from where the Contractor's Head Office is located.

There is a far easier way for even large-sized Contractors to handle the recruitment process, and which will leave the HR Department's personnel free to deal with all other important HR matters more effectively. That way is to have long-term arrangements with reputable recruitment agencies, who will then do all the pre-screening work of the candidates and compile shortlists of the most suitable people for the Contractor's consideration. The major benefit of such arrangements with outside recruitment agencies is that special relationships can be forged so that the Contractor's requirements are prioritised, thus cutting down the time required to place people in position. Since the recruitment agencies are full-time in the business of recruiting people, their databases are very likely to be up-to-date whenever the Contractor has a requirement to recruit somebody new. In the long-term, this approach is likely to be far more satisfactory than the Contractor attempting to do everything on its own through its HR Department. Of course, care must be taken with the wording of the contractual provisions contained in the agreements signed with the recruitment agencies, particularly in respect of:

(i)   what constitutes the introduction of a candidate (especially if the introduction is made via two agencies at the same time, or if the candidate declines then comes back months later of his/her own accord),

(ii) what steps need to be taken if a new employee fails to do the job properly, etc., and

(iii) how much discount will be available where many placements are made through the same agency over a given period.

Another problem area is how Disputes are to be handled, and where the legal jurisdiction for settling Disputes will be. The HR Manager should always ensure that the Corporate Legal Department (CLD) is involved in the formulation of such agreements, since the CLD will ultimately be responsible for handling the matter if a conflict arises that might lead to litigation. Involving the CLD is always the best way for the HR Manager to protect himself/herself if the contractual provisions are ultimately found to be not as good as the Contractor had been hoping for.

6.22.4   Emergency Replacement for Key Personnel

For any large Contractor, it is inevitable that some of the employees will be found lacking in their job performance capabilities. This can be for a variety of reasons, such as ill heath, family problems that cause the employee too much stress, incompetence, and even drug abuse or alcohol abuse issues. The HR Manager should always be prepared for this eventuality, and especially so for key personnel. It is therefore always best to identify others within the organisation who could possibly temporarily replace key personnel in distress if something untoward were to happen, at least until such time as a permanent replacement can be found. More than one potential replacement needs to be identified, because good people are generally fully engaged in the field and therefore cannot be moved easily.

The above risk reduction technique will require compiling an adequate interlinked/cross-referenced database that needs to be maintained and kept up-to-date at all times, as and when new people are taken on or existing employees depart. The records in any such database need to include fields to show where each potential key replacement staff member is currently working, and how long the commitment to be in position actually is. Such preparation should help the Contractor to avoid the situation where an outgoing key worker that has to be replaced urgently is replaced hurriedly by somebody from outside who subsequently proves to be unsuitable.

6.23   The Administration Department

6.23.1   Primary Responsibilities

Many EPC Projects will probably not need a complete Administration Department set-up. Instead, the Project will more usually rely to a great extent on the staff in the Contractor's main office (corporate) Administration Department for the majority of the administration inputs, supported on the Site by one or more Administration Officers. Those Administration Officers will usually be entirely responsible for handling local administration matters, and they will also act as the go-between for the Project with the main office in respect of all other administration matters. The staff members in the main office's Administration Department tend to work in the background, and the only time anybody is really aware of their input is when administration matters start going wrong. However, the work of the corporate Administration Department is vital for keeping matters ticking over smoothly for the Contractor.

The corporate Administration Department Manager's role is to manage, organise, direct and monitor the personnel working in the main office's Administration Department, as well as the Site-based administration support staff, ensuring that, collectively, they carry out the following primary responsibilities satisfactorily:

1. Obtaining visas and work/residence permits for all personnel when the work is in a foreign country, both for the main office and the Site-based personnel.

2. Arranging for air travel and ground transportation of the personnel as and when necessary, including the provision/coordination of security details in locations where personal security is at risk.

3. Arranging suitable accommodation and boarding facilities for all personnel working in the main office and also those involved in the Project at the Site, including Vendors' and Subcontractors' personnel.

4. Arranging for the provision of sufficient main office facilities as well as Site-based facilities, including the provision of adequate communication and entertainment systems.

5. Procuring and delivering/installing all office equipment, furniture, supplies and utility services for efficient running of the whole of the Contractor's business in whatever locations.

6.23.2   Providing Back-Up Support to Site Quickly

The biggest risk for the Administration Department Manager is if the Site-based support staff members fail to perform well or if the key Administration Officer falls sick suddenly, since it often takes a long time to obtain visas, work permits and/or residence permits for replacement personnel. The best way to avoid this situation happening is to place a trusted long-term staff member in position on the Site, supported by competent personnel who could take over at short notice and manage sufficiently well until a replacement for the key person becomes available.

It is not generally too difficult for the main office to secure replacement services for itself if the original supplier/provider lets the team down and fails to provide a good service, but the situation for a major Project in an overseas location may be completely different. It is therefore prudent not to put all the eggs into one basket when it comes to overseas local suppliers/providers and, instead, split the work between at least two companies, so that a backup is available at all times if one entity becomes unreliable. This is particularly true for security personnel details, since movement will become severely restricted (or even impossible) if protection is not immediately available. When transportation is needed for the Contractor's staff in hostile environments but is not available, it can become difficult for some people to remain calm about the situation. Such panic situations must be avoided as much as possible by having adequate contingency plans in place, just in case 'Plan A' fails.

6.24 The Financial Management Team

6.24.1 Primary Responsibilities of Finance Manager

The Finance Manager is usually a corporate, Head Office based role. He/she will therefore have many corporate functions to perform that are not relevant to what this book is about. The following therefore sets out only the primary responsibilities of the corporate Finance Manager in respect of providing financial support to a Project.

Cash flow is of vital importance to every Project, since there will be many other companies on which the Contractor will be relying to supply materials, goods and equipment, undertake work and provide services. However, most Projects do not demonstrate a neutral cash flow for the Contractor, despite a number of the world's largest Employer bodies stating that such a situation is what they aim for by (i) not withholding 'retention money' and (ii) ensuring a fair 'progress payment' system. As a result, they think that they do not need to give Contractors any advance payment for the Project. I consider that the foregoing views of the Employers do not hold water, because of the following:

1. The Contractor will almost certainly be required to provide a Performance Bond, most likely equal to 10% of the Contract Price, which the Contractor should consider is in lieu of having retention money deducted from interim progress payments.

2. Some International Oil Companies and Employers also insist on deducting 10% retention money from interim payments due to the Contractor in addition to requiring a Performance Bond. And some may even demand a Parent Company Guarantee on top of that.

3. There will be many instances where the Employer will consider that payment is not justified unless something tangible is delivered first. For example, mobilisation of personnel for the Engineering services will not be considered for payment until certain minimum documentation 'Deliverables' have been approved by the Employer's Team (not just provided to the Employer's Team for review purposes). These will usually comprise the key procedures to be followed for the Project's implementation. Even if the Contractor is able to complete such Deliverables within four weeks, it will probably be a further four weeks before approval of those documents is given.

4. Most Vendors will require substantial upfront payment before committing to commencing manufacture of goods/equipment. They will almost certainly require full payment before the products are handed over to the Contractor at the agreed point of delivery. However, the Employer will generally refuse to reimburse such upfront payment amounts to the Contractor, and most certainly will not pay 100% of the cost of the goods/equipment until they have been delivered to the Site. Even then, many Employers will hold back a certain amount of money until satisfactory commissioning of equipment has been completed.

5. Added to the above financial burdens and cash flow problems, the payment approval process itself will probably take at least two weeks for each invoice the Contractor raises, and the period for honouring payment of invoices will likely be a minimum of 60 days. All-in-all therefore, the Contractor may be faced with carrying the cash flow requirements of the Project for a minimum of 20 weeks. This means that, without an advance payment on the table, talk of the Contractor being offered a neutral cash flow payment arrangement can generally be seen to be complete nonsense.

The Finance Manager is the person responsible for making sure that the funding requirements to provide for the cash flow needs of the whole of the Contractor's business are properly calculated and conveyed to the Contractor's Corporate Management Team and the Board of Directors, so that approval to raise whatever additional funds are required can be obtained. Despite the foregoing fact that neutral cash flow cannot be achieved for most of the Contractor's Projects, the Finance Manager must ensure that every individual Project has the necessary finances made available to it for the entire duration of each respective Project's implementation activities. This requires the Finance Manager to be fed with all the correct data from each Department, and each Project, to be able to complete this exercise meaningfully. To ensure that is achieved, the Finance Manager needs the support of a Cost Management Team for each Project, with each such team led by a Budget Controller. On each respective Project, the Budget Controller is required to report to both the Project Manager and the Finance Manager about the Project's ongoing financial status. The corporate Finance Manager's biggest risk is that the financial data and information coming from the Projects do not reflect the true situation. I suspect that it is highly likely that this is what happened to the team responsible for managing Carillion plc's[30] financial well-being.

6.24.2 The Budget Controller

The budgets and cash flow requirements for the main office Departments will generally not be too difficult to ascertain, because the staffing levels will be fairly static most of the time. However, the cash flow requirements for the entirety of the Projects is another matter completely. What each Project therefore requires is for a Budget Controller to be appointed, whose task will be to keep track of all expected and actual expenditures, as well as expected and actual incomes, so that the Project's cash flow requirements can be ascertained with a great degree of accuracy for reporting to the Finance Manager. If there is no Budget Controller, the Finance Manager's cash flow projections may be seriously at risk because of the Finance Manager not receiving bad news in good time. The result

---

30 Ibid., Section 1.1.

of that could be that there will be insufficient liquid funds to finance the day-to-day operations of the Contractor's principal offices as well as those of all the Projects.

The Budget Controller's task is to monitor each and every major item of expenditure on the Project, and to shout loudly to both the Finance Manager and the Project Controls Manager when it looks as though expenditure for any item is going off course. This will enable early action to be taken in an effort to stay within the Project Budget, even if it means that some of the contingency fund has to be used up.

The Variations occurring with Purchase Orders are usually far easier to spot, because most Vendors refuse to supply anything extra unless an official instruction is made in writing. On the other hand, Subcontractors have a habit of doing additional work based on informal instructions, then submitting the request for payment much later down the road. This is why it is necessary to have Quantity Surveyors and/or Cost Engineers on board, whose task it is to keep the Budget Controller fully informed about the financial status of all the Subcontracts, and to flag up the situation ahead of time whenever it seems that large extra costs will be involved. Failure to keep on top of the detailed expenditure situation could result in the expected profit being eaten up through myriad small increases across the various Purchase Orders and Subcontracts, without anybody being aware of it until it is too late.

6.24.3 Controlling Purchases

In regard to cost control and expenditure management, the following words, often attributed to Peter Drucker (management expert), should be constantly at the forefront of the mind: 'What gets measured gets managed'.[31] That statement is never more true than in the construction industry where, on large capital Projects, 50% of the expenditure is often on materials, goods and equipment. Another 20% or more is generally spent on the costs of the manpower and supervisory inputs necessary to install those items to form the completed work. If more attention was paid to measuring those 70% of items, then that would go a long way towards helping to avoid cost over-runs. However, I have seen too many Contractors allow the Procurement Department to order items with little input from people experienced in the totality of what is needed to complete a Project. The result is that 'top-up' orders are continually being placed at exorbitant unit rates, and budgets consequently being blown apart. Perhaps this would be more understandable for local Contractors in some developing countries, but I have seen it happen often in international construction companies too.

Further, if Corporate Management Teams and Boards of Directors could be made to fully understand how strong the temptations are for Procurement Department personnel if purchasing controls are lax, then far greater control over purchases and subcontracting would soon become the order of the day. Most Procurement Managers are under intense pressure to get the Purchase Orders placed as quickly as possible, and so they do not have the luxury of time to police the work in their Department to the depth that is most probably needed. It is therefore imperative, in my opinion, for an in-house audit team to

---

31 Wikipedia. Drucker, P. Described as the founder of modern management; https://en.wikipedia.org/wiki/Peter_Drucker (accessed 5 March 2018).

be regularly vetting the activities of the Procurement Department, as well as the work of those administering the financial aspects of subcontracted work, as a matter of course.

To undertake management of the financial business of the Contractor properly, and depending on the way the Subcontracts have been set up, the Project will require a team of Cost Engineers and/or Quantity Surveyors (the Cost Management Department) to conduct the day-to-day cost control activities. However, many EPC Projects do not include an adequate number of such costing/measurement personnel, and I have been given the following as one of the reasons for that: 'The EPC Contract is lump-sum, so there are no measured quantities required'. The fact that the Contractor's bid price was submitted on a lump-sum basis (and that the Contract Price too was agreed on that basis) does not mean that detailed quantities will not be required by the Contractor. It needs to be remembered that quantities will be required for all the bulk materials so that accurate Purchase Orders can be placed, and they will also be needed to call bids for work where the Subcontractors are to be paid on a measured work basis (not just on completion of the Subcontractors' work but monthly too, for interim payment purposes).

6.24.4 Keeping Track of Expenditure

Another big risk faced by the Finance Manager is that the Project's bid price may have been severely under-cut, but which may not become apparent until much later in the day. Early measurement of the finally required quantities by personnel skilled in such activity (and who can also accurately price the measured items, such as qualified Quantity Surveyors and Cost Engineers can do) will give the Finance Team the opportunity to take mitigating action in order to reduce the costs of any elements of work found to be exceeding the budget. Establishing that an element of work has exceeded the budget only after the bids are in will inevitably delay the work while the remedial plans are being put into place. It is therefore incumbent on the Finance Manager to determine at the bidding stage what staffing levels are required in his/her financial support team beyond just the Budget Controller. If the bid price did not allow for an adequately sized team, then trying to get the right staffing levels after the Contract has been awarded will almost certainly meet with great resistance from the Contractor's Corporate Management Team.

6.25 The Compliance Team

Nowadays, in order to be eligible to work for international Employer bodies, it is imperative for the Contractor to have in place adequate policies and procedures that attempt to ensure that the Contractor's personnel (and those of Vendors and Subcontractors too) are legally and ethically compliant in all respects. This is not something that happens naturally but requires a team of personnel assigned to the task, whose members are dedicated to conveying the principles of ethical compliance throughout the organisation, and who check on a regular basis that international ethical compliance requirements are being diligently followed and applied.

The Compliance Team is not usually part of the Project Implementation Team, but it is required to interact with all those engaged on the Project to ensure that proper training in respect of ethical compliance is given to everybody. This is essential because if any of the Contractor's personnel is found guilty of offences such as engaging in bribery or corruption, the Contractor will be held liable too, unless it can be demonstrated unequivocally that the Contractor has taken all reasonable steps to convey the ethical compliance requirements to its employees. That can only be done by way of formal training, with adequate supporting policies and procedures in place. It also needs to be demonstrated that regular audits are conducted that allow ethical compliance to be monitored by the Contractor in an effective manner.

If any major Employer considers that the Contractor is paying lip service only to internationally required ethical compliance legislation, then there will be a severe risk that the Contractor will not be invited to bid for future Projects for that Employer. The rationale behind this is very simple: the Employer would be just as liable as the Contractor if any of the Contractor's personnel violated ethical compliance regulations on the Employer's Project in the situation where it was apparent that the Contractor did not take such compliance seriously enough. Most Employers will not take that risk upon themselves. It is therefore very important that adequate time allowance is built into the Project Schedule to accommodate the time it will take for the training of workers in regard to ethical compliance requirements. Doing so, and then ensuring that such compliance training is fully implemented, should serve to eradicate the risk that an Employer will strike the Contractor off its bid list for future Projects, even if something untoward in respect of ethical compliance violations did occur.

One of the areas where a Project can often find itself being on the delivery end of an ethical non-compliance report issued from the Employer's side at the commencement of a Project is in respect of not re-prequalifying long-standing Vendors and Subcontractors where necessary to do so. Another area of non-compliance would be not conducting regular audits on the performance of Vendors and Subcontractors, or not undertaking an assessment of their business/work ethics. Later, this situation could develop into the Contractor receiving a court summons from an international regulatory authority. The problem is that a certain amount of complacency sets in with Vendors and Subcontractors that a Contractor has been dealing with for years, and a level of trust is developed that is not generally questioned. However, that is precisely where trouble spots develop. A good example of that was where, not so long ago, a well-known IOC had been unaware that, for many years, its trusted Subcontractor responsible for transportation of goods in a foreign country was bribing officials in many of the provinces through which the heavy goods vehicles passed. Apparently, the IOC had failed to ensure that the Subcontractor had been properly informed/trained about the need to run an ethical business. The fines levied against the IOC were substantial.

The problem for the Contractor is that the responsibility for keeping on top of ethical compliance checking of third parties involves many different Departments:

(i) the corporate Proposal and Procurement Departments at the bidding stage,

(ii) the Project Procurement Department at the purchasing stage,

(iii) the Project Contract Department and/or Subcontracts Management Department at the subletting and close-out stages, and

(iv) all the various Managers and Sub-Managers engaged during the construction stage responsible for the supervision of Vendor and Subcontractor activities, where many of the management personnel involved might not be aware of the ethical compliance requirements or simply assume that somebody else is dealing with such issues.

The problem is exacerbated when the work is being done remotely from the Site where the Project is being implemented. It also needs to be remembered that debarment from carrying out one funder's Projects (such as World Bank) can also end up as cross-debarment across many others (such as Asian Development Bank, European Bank for Reconstruction and Development, Inter-American Development Bank, etc.).

The only way to satisfactorily reduce the risk of third parties compromising the Contractor's business is for the Contractor to establish and maintain a fully-staffed corporate Compliance Department, with properly trained personnel on board, where the main functions of the Department should be to ensure that:

(i) all necessary prequalification (or re-prequalification) work and performance monitoring for the Vendors and Subcontractors utilised by the Contractor is rigorously performed,

(ii) ethics training is given by the Contractor both to the management staff of those other entities and their employees too, and

(iii) constant monitoring is conducted of the business activities of those other entities to ensure that they operate ethically and have not subsequently ended up on the sanctions lists of compliance authorities.

For that work to be undertaken thoroughly, the Contractor's Chief Compliance Officer must be given suitable authority to demand answers from any and all of the Contractor's personnel. This is because one of the classic ways to try to cover up where 'dirty business' is going on is to avoid answering audit questions properly. Such non-cooperation with the Compliance Team should be met with appropriate disciplinary penalties being applied. In that way, everybody will be made aware of the need to work ethically at all times and not make life harder than it should be for the Compliance Team members to establish if problems exist that could damage the Contractor's reputation for conducting clean business.

6.26 The Legal Department

6.26.1 Utilising External Legal Team

Unless the Contractor is huge and engaged on a large number of mega-sized Projects, a Legal Department is not usually a necessity for an EPC Contractor. However, just occasionally, the Contractor may need to call in legal experts if a major Dispute arises. The reality is that litigation to resolve Disputes costs a great deal of money, takes up a lot of time for the Contractor's Senior Managers and can tend to sour business relationships over the longer term, not just with the Employer but with the rest of the business community. Further, most Employers are very wary of doing business with any Contractor who has a reputation for engaging in or threatening litigation regularly.

For this reason, my advice is for Contractors to utilise outside lawyers only when really necessary. However, to make sure that such advice will be available when required, it would be a good idea for the Contractor to sign a long-term Service Agreement on a retainer basis. This has the advantage that the law firm's senior personnel will become very familiar over time with the workings of the Contractor's own organisation. They will then be able to jump in at a moment's notice to help out in times of trouble. This is very useful in countries where the legal system is not as formally (or cleanly) run as it is in the Western world.

In parallel with the above approach to limiting the engagement of outside lawyers on a Project, the Contractor's Corporate Management Team should instruct all those responsible for administering Contracts, Subcontracts and Purchase Orders that the Contractor's policy is for those staff members to act as facilitators at all times, and not become troublemakers. Wherever reasonableness is not being reciprocated by the other party, the Contractor's administrators should avoid personal conflict as much as possible and elevate the problem to the Contractor's Corporate Management Team to handle. I offer this advice because I have observed that issuing a threat about the dire consequences that could occur if the other party does not agree with the Contractor's solution to a sticky problem only tends to make matters worse, not better. Thus, if such things need to be mentioned, I feel that it is far better that it is done as a balanced discussion (i.e. not done in the heat of the moment) between the Directors of the disputing parties alone, not involving those directly engaged in the actual work itself.

6.26.2   Utilising Dispute Adjudication Boards

Having said all this, I have also observed that the 'FIDIC' concept of setting up a permanent Dispute Adjudication Board (DAB) to help avoid tricky situations turning into ugly Disputes works very well. I consider that is primarily because it reduces the chances of personality clashes getting in the way of reaching a mutually agreeable settlement. A DAB is a low-cost Alternative Dispute Resolution technique that the Contractor must ensure is incorporated into the contractual provisions before Contract signing. It will rarely be agreed to by the Employer after the Contract has been signed if it was not provided for from the outset. Having said that, I have seen a similar sort of arrangement agreed to under those circumstances just once (where, ultimately, it proved to be very successful).

To be truly effective, the members appointed to the DAB should be knowledgeable experts, who would be required to work together with the Project's key participants to obtain a solid grounding/understanding of the Project before launching into the task of resolving Disputes. The main purpose of utilising a DAB is to resolve each Dispute as quickly as possible with the minimum of fuss and the maximum amount of confidentiality. This helps to avoid acrimony arising between the disputing parties, and gives each of them the earliest possible certainty as to where they stand, so that they can then move on with their respective businesses in confidence.

Chapter 7

Reducing Joint-Venture/Consortium Risks

7.1 Joint Venture Versus Consortium

It is not unusual to find that mega Projects are undertaken by a group of companies that have formed either a Joint Venture (JV) partnership or have agreed to operate as a Consortium. There are some similarities between the two contractual arrangements, but Consortiums tend to have less acceptance with Employers. For that reason, Consortiums face more administration-based difficulties than JVs do. The more usual problem areas with Consortiums are: (i) the provision of satisfactory Performance Bonds and insurances, and (ii) difficulties in setting up bank accounts for payment purposes that are acceptable to the Employer. This is mainly because of the more casual (less formal) relationship that Consortiums operate under compared with JVs. However, even JVs can have problems with providing a single satisfactory Performance Bond, and I have experienced that situation. The Employer's response was to terminate the Contract even though the on-Site work had commenced. The loss of face and money was significant for both Contractors involved.

To simplify matters hereinafter, from this point on I will refer to such cooperative joint working arrangements only as JVs, on the basis that all such references should be understood to refer to Consortiums also.

7.2 JV Considerations

The cooperative working under a JV arrangement more often than not often extends all the way from the expression of interest and bidding stage right through to handover of the completed Project. Sometimes it even extends to operating and maintaining the completed facility. JVs are very often the preferred route for general contractors to take when specialised, patented equipment is involved on a large scale that only others can provide (such as catalytic cracking units for a refinery). However, a specialist party's involvement/share in the JV may be significantly smaller than that of the other party or parties handling the remainder of the work, simply because the cost of the specialist work is small compared with the value of the whole Project.

Managing JVs is far more complex than managing a single business entity, and therefore they often run into trouble. A 2014 *Construction News* article reported that construction consultant EC Harris's recent research ('Global Construction Disputes: Getting the Basics Right') had found that, in the UK, 'A third of construction joint ventures now end in dispute'. The same annual survey three years earlier had found that 'one in four joint ventures ended sourly'.[1] When JVs do go wrong, the well-being of the Project can suffer tremendously and end in abject failure. A lot of effort must therefore go into setting up the JV so that it stands the best possible chances of working successfully.

Although it almost goes without saying, the JV must be formed amongst entities that have sound technical and financial capability. Anything less would almost certainly be asking for trouble, with the better/financially stronger entity being required to pick up the tab for its partner's failures. A worthwhile formal JV agreement should also be drawn up. That document should deal fully with all aspects of the intended relationship, including what should happen if things go wrong. However, I have seen some very weak JV agreements that were found to have left a lot to be desired when things eventually went wrong. Such situations generally occurred because the JV was formed between entities where the key personnel in each organisation had been friends or close acquaintances for many years. The problem with a loose JV agreement is that the people managing the actual work may not be aware of such close relationships and thus may not have the same ethos or objectives of the people who drew up the agreement.

Insisting that a solid set of operating rules should be drawn up for the JV should not be seen as distrust on the part of the person doing the insisting. It should simply be regarded as sound common sense. The JV agreement should clearly state the aims of the members, so that there is no misunderstanding as to what needs to be achieved and what each entity will be contributing to the successful completion of the undertaking for which the JV is being formed. Those objectives can then be conveyed to all those responsible for handling the JV's daily activities.

In short, the primary aim of the JV agreement must be to commit all members of the JV to fulfilling all the requirements of the Employer as embodied in the Project's Contract Documents. Provided that each JV member does its bit to complete its work properly and pay attention to reducing the risks highlighted in the earlier chapters of this book, then the chances of the JV completing the Project successfully should be enhanced significantly.

7.3 Setting Up a JV Steering Committee

To ensure that there is a permanent forum for discussing problems amongst JV partners, a Steering Committee should be formed, comprising senior representatives from each of the companies involved. The Steering Committee should then arrange to hold regularly scheduled meetings. Where more than two companies form the JV, it is recommended to keep the number of participants down to just one representative from each

---

1 *Construction News* (27 May 2014). A third of JVs end in dispute – British contractors are getting worse at joint ventures, new research suggests. www.theconstructionindex.co.uk/news/view/a-third-of-jvs-end-in-dispute (accessed 13 February 2018).

company. This is in order for such meetings to be effective and for quicker decisions to be reached. Too many people at a meeting will inevitably make it difficult to reach consensus opinions quickly. If four companies are involved in a JV, then one member from each company would still mean that four people would be present at each meeting. That should be more than enough to be able to reach sound decisions, bearing in mind the required seniority of the people involved. However, for a two-party JV, it would be satisfactory for there to be two representatives from each company.

If the place where the JV Steering Committee meetings are held is also alternated between the various overseas Head Offices of the companies (so as to encourage stronger bonding between the companies involved), it would mean that, to attend the monthly meetings, each month three people would be required to fly internationally where four companies comprise the JV. That would not be a small expense, and it would also eat up a lot of time in travelling. Adding additional numbers of people would therefore generally be wasteful expenditure.

Ideally, once the Project activities have reached the on-Site stage, the accommodation for those attending the JV Steering Committee meetings should be arranged in the Base Camp or Fly Camp (as applicable), thereby saving on hotel costs. Such attempts at genuine cost-savings do not usually go unnoticed and will generally be appreciated by the Contractor's Project Management Team. Such cost saving efforts will also set the right tone for all other Project participants. Added to that, the level of bonding and camaraderie achieved by the top Managers taking part in after-hours social activities at the Site can go a long way towards lifting on-Site morale, especially if the Project is facing unexpected difficulties. All these little things can contribute to a better bottom line for the benefit of all the entities comprising the JV.

Of course, it almost goes without saying that the JV representatives on the JV Steering Committee should be carefully chosen to ensure that facilitators are selected over more disputatious people. However, I have seen situations where that patently was not the case, so that the resultant personality clashes wrecked the harmony of the JV beyond repair. To be able to deal effectively with such disastrous situations, the JV agreement should unequivocally state that, where it is found necessary to recommend the replacement of a disruptive representative, then a majority vote by the other members to have that person replaced will suffice to give effect to such recommendation. In the situation where a majority decision is not feasible, then the JV agreement should make provision for an independent person to have a casting vote. After all, it would be far better to put one person's nose out of joint than for the whole of the JV to become incapacitated.

7.4 Objectives of JV Steering Committee

Ideally, the JV Steering Committee members should be the Project Directors for the various JV partners (with any Deputy Project Directors participating in two-party JVs or, in a JV with more than two parties, acting whenever necessary as the stand-ins for the Project Directors). Such members should not generally form part of the full-time job management team involved in the day-to-day activities on the Project (except perhaps where the workload of any particular JV partner is large enough to warrant appointing a full-time on-Site Project Director). This is because the primary purposes of the JV Steering Committee are as follows:

(i) to bring the JV partners together to discuss and resolve all matters affecting the activities of the JV itself (as distinct from the daily goings on of the Project), and to foster good relations between the various companies participating in the JV;

(ii) to compare notes about the status and progress of the Project, and to raise concerns if it seems that anything is not going according to plan and could thereby jeopardise the Project's objectives;

(iii) to perform a top-level checks-and-balances function on all major issues in order to ensure that matters are being attended to properly (such as by reviewing the health, safety and environmental compliance statistics and comparing the results/outputs of such activities against the Key Performance Indicators), and

(iv) to put in place appropriate measures to rectify any areas where the results/outputs do not meet the required standards.

7.5 JV Members are Partners

The biggest problem I have observed with JVs is when it comes to deciding which member will be the Lead Partner for the JV. This usually arises because of the egos of the different Boards of Directors, and sometimes because each entity involved considers that they are better qualified than the other members of the JV. What I sometimes saw being forgotten was that members of a JV should treat all other members as partners. This is despite the reality that the entity responsible for undertaking the most critical element of the work, when considering both the technical aspects and the financial burden, should be appointed as the JV's Lead Partner. Having said that, I have observed that this is not so easy to decide in the situation where the Engineering work is done by one entity and the Procurement work by another, with the Construction activities being undertaken by all the entities involved. However, my advice is that the Lead Partner should simply be seen as the spokesperson for the JV, and all its members should be viewed as partners who have an equal say when it comes to discussing policy internally. Quite obviously, there will need to be a mechanism for overcoming impasses on critical issues, but that should still respect the principle of all members being treated as equal partners.

I have, however, seen examples of where the Lead Partner treated the other JV members almost as Subcontractors, with dire consequences. Such situations were completely avoidable and entirely unnecessary, in my opinion, and were sometimes made worse by the means of communication adopted by the Lead Partner. I am referring to where formal letter writing from the Lead Partner took the place of cordial meetings and friendly dialogue to resolve problems and agree on the action to be taken. In almost all cases this led to an exchange of highly legal/contractual letters that did not improve the atmosphere between the parties. The most successful JVs I saw relied more on face-to-face meetings conducted in a relaxed manner to resolve any differences/problems, operating as if the members were friends, not adversaries. Emails, using friendly language, were then used to confirm the agreements reached. The knock-on effects of a failed JV to the Lead Partner, such as loss of face in the wider business community and the very large additional costs of overcoming the departure of a valuable partner, should be carefully considered by the Lead Partner. My advice is that every effort should be made by all JV partners to forge a very close and friendly relationship amongst their respective management teams. That can be achieved by all entities being open and frank with each other, and communicating well, and in a friendly manner, on all important issues.

I will close this chapter by stating that I have been fortunate enough to work on a number of Projects where JVs operated very successfully, and thus I was able to see very clearly that the success was due to the exceptional calibre of the people comprising the JV Steering Committee. On the other hand, the failures of the JVs I observed were almost always due to the presence of management personnel in the JV Steering Committees who were difficult to deal with. Often, such personnel displayed marked negative tendencies to the point of being naysayers; how they managed to be appointed in the first place was beyond me.

Based on my experience, I agree that some healthy scepticism is essential when key issues need to be resolved, in order to counter the possible wishful thinking that sometimes accompanies the complacency that sets in when things are going well. However, in general I prefer (and therefore recommend) a questioning approach to resolving differences of opinion and getting to the core of the problem, rather than adopting an aggressive approach to seeking answers to difficult problems. In any confrontation it is all too easy to allow ego to jump in and dominate a discussion, with practicality then being thrown out of the window. I have therefore concluded that choosing the right ambassador to represent each entity in a JV is one of the most critical elements in reducing the risks of the JV failing. Even then, I am sure from what I have seen that it will still be a constant battle for a JV to deliver a truly successful Project for all the JV participants.

Chapter 8

Claims Management Risks and Problems

8.1 Relying on Claims to Achieve Profitability

In the past, it was not unknown for less scrupulous Contractors to rely upon their Claims prowess to secure their profit. Their ploy was to submit heavily qualified proposals against a low bid price that, on the face of it, seemed beneficial to the Employers. However, at the end of the day, it cost those Employers far more than they had bargained for. To a limited extent that ploy is still in use today, but usually only amongst companies that operate in specialist niches where a lot of expertise is required to understand the nuances of the bid qualifications. Examples of such specialist niches, where I have found highly litigious and very competent claims-orientated personnel, are marine works specialists, proprietary major equipment suppliers, and specialist technology suppliers such as SCADA (Supervisory Control and Data Acquisition) design and installation companies. I have too often found that the specialists for such companies attempt to bamboozle opposing arguments by employing a great deal of technical jargon that few outside of their specialisation understand well.

In my opinion, it is not at all wise for the ordinary general Contractor to have to rely regularly on Claims on EPC Projects to take the company's fortunes from misery to happiness (as per the Micawber Principle).[1] Sadly however, far too many EPC Projects nowadays seem to have trouble making the profit that was anticipated. Worse still, many of them significantly overrun their original completion dates. That means that making an extension of time request very often becomes an absolute must, if for no other reason than to obviate the need to pay Liquidated Damages for delayed completion. The result of this all-too-common situation is that last-minute Claims are then often required to be prepared in haste, and submitted in the hope that they will receive a fair review. However, the playing field for resolving Claims much more resembles a minefield than anything else, with all sorts of vested interests mitigating against the Contractor's Claim being successful.

---

1 Wilkins Micawber, the ever-optimistic character in the novel *David Copperfield*, by Charles Dickens.

8.2 Factors Legislating Against a Claim's Success

There are many factors legislating against the Contractor's Claim being successful. First, when contemplating submitting a Claim, it needs to be remembered that Claims will always be resisted by the Employer's Team, and particularly so by a Project Management Consultant (PMC). Largely this is because Claims have the potential to reflect badly on the ability of the PMC team members to manage the Project well. It therefore does not matter how friendly the relationship is between the Contractor and the PMC because, as soon as the Contractor submits a significant Claim, the hackles of the PMC will automatically rise, and the Claim will be thoroughly scrutinised to find any possible cause for the PMC to reject it. Further, many Contracts contain a provision that invalidates Claims from the Contractor that are not submitted within a specified time-frame. Although there are different opinions in different jurisdictions about the validity of such 'time bar' provisions, it is always best to notify the Employer's Team about any potential Variation the moment such an issue is noticed (clearly stating that it appears to be a potential Claim). Such notification must then be followed up with detailed substantiation to ensure that a worthwhile Claim has been lodged. Failure to give timely notice of the existence of a potential Claim can scupper the chances of that Claim being successful before it has even been compiled. Failure to provide the substantiating backup data in good time can have the same adverse effect.

Second, it must be borne in mind that many Claims fall down simply because the Contractor's Team has not kept adequate records of the impacts of the individual issues that have caused additional costs for the Contractor and delay to the Project's progress. When the Contractor's Team finds itself in the situation of having inadequate supporting information for individual Claims, it very often means that the cost impacts of each separate cause of delay cannot be gauged properly. This is often compounded by the lack of a valid Project Baseline Schedule and/or the lack of valid updates thereto. The usual tactic then employed is for the Contractor to submit what is known as a 'global' Claim, which is generally regarded as a very weak methodology to adopt. Such Claims have a hard time in achieving the success level hoped for by the Contractor, so should be avoided wherever possible. This because, as Kennedys Law LLP's article shows, there are often valid lines of defence for the Employer against global Claims.[2]

To avoid the problems outlined above, the Project Manager must ensure that all its Managers are fully briefed at the commencement of the Project as to what to do to record the effects of unexpected changes that occur. It also needs to be stressed to the Managers that all the support details of such changes must be communicated to the appropriate person responsible for handling such a Claim as quickly as possible. Expecting the Contract Administration Manager (CAM) to trawl through the documents for relevant data late in the day often proves to be a laborious task that takes up a huge amount of time. This is owing to the large volume of data often involved (i.e. letters, minutes of meetings, progress reports, requests for information and clarification, comments emanating

---

2 Butler C., Carter P. and Cooper L. (19 January 2017). Global claims: a brave new world? http://www .kennedyslaw.com/article/global-claims-new-world (accessed 10 May 2018).

from the reviews of the Contractor's Deliverables, design review meetings, Project/site diaries and document transmittal data, etc.).

Whatever the CAM produces under his/her own steam can never be as thorough a compilation as if the Managers dealing with the individual problems had themselves bothered to pull all the relevant data together. Therefore, leaving everything for the CAM to do from scratch cannot possibly help the Contractor to submit a fully documented Claim in a timely manner. For this reason, I recommend that the agenda for each monthly Project meeting should contain an item for potential Claims to be raised as a discussion point, thereby giving the CAM the opportunity to hammer home the importance of good record keeping and timely communication of any pending problems.

8.3   Key Ingredients for Worthwhile Claims

In view of my personal experience of being brought in belatedly to prepare Claims, I feel that is worth stressing that, to ensure the success of a Claim, there are three key ingredients that the Contractor must be in possession of before launching it:

relevant facts, more relevant acts, and even more relevant facts.

Without an adequate array of relevant facts, the Contractor will not be in a dominant position when it comes to persuading others about the strength and worthiness of its Claim. On the contrary, launching a Claim with inadequate relevant facts will give the other party a great opportunity to rip a Titanic-sized hole through that Claim as soon as it has been launched. A great deal of attention should therefore be paid to ensuring that all Managers are fully aware of the need to compile the support papers for each and every Claim from the moment the issue is first spotted. A copy of all such data should be saved in a confidential folder on the Project's server that can be easily accessed by those who will be responsible for having a hand in processing any aspect of the Claim.

8.4   Proving Excusable Delay

One of the major problems when lodging a Claim is providing adequate data to substantiate the extension of time necessary to complete the work. The primary cause of that problem will normally be that no attempt has been made by the Contractor to update the Project Schedule on a monthly basis to reflect actual progress. All attempts by the Contractor to retrospectively update the Schedule will be seen by the Employer's Team as 'fudging of the facts' to support the Contractor's case, and done with the express intention of concealing any of the Contractor's delays.

The negative attitude of the Employer's Team towards any structured attempt at conducting delay analysis is hardly surprising considering that, even today, there is still no consensus of opinion as to how a fair extension of time is to be established. This was evidenced most clearly by 'The Great Delay Analysis Debate' held at King's College London on 18 October 2005,[3] where papers were presented to debate the motion: 'This house

---

3 Presented by the Society of Construction Law, in association with the Centre of Construction Law & Management.

considers that the time impact method is the most appropriate for the analysis of delay in construction disputes'. Although that motion was rejected, those papers were later published for general distribution.[4] As part of the debate, four different approaches to delay analysis had been tabled. When the audience (of around 300 people, including some of the UK's leading experts in delay analysis) was asked in turn which of the four methods they favoured, none could command a clear majority and each had a committed band of supporters.[5]

It can involve a lot of time and money to retrospectively build the Project Schedule and subject it to detailed delay analysis techniques such as the time impact method. All such efforts can be a complete waste if the Employer's Team rejects the output out of hand. If the Contractor has also failed to flag up the delay problems properly in the monthly reports, this will only add weight to the correctness of the outright dismissal of the Contractor's submission by the Employer's Team. I therefore cannot stress enough how much easier and effective it would be to keep the Project Schedule up to date properly, right from the outset of the Project, so that it will be there, ready and waiting, when it is needed to support the Contractor's extension of time application. The fact that the Contractor will need to make such a submission at some point in the life of a Project is almost inevitable. I can remember only a very few of the many hundreds of Projects I have been involved in where delayed completion was not evidenced. Not one of those more successful Projects was anywhere near the size of a mega Project, and certainly not EPC in nature.

8.5 Key Components of Successful Claims

In closing this topic, I would just like to stress that I consider that the success of any Claim is dependent on:

   (i) there being solid contractual grounds enabling the Claim to be made,

   (ii) the storyline behind the Claim being indisputable,

   (iii) the Claim being raised in a timely manner,

   (iv) the supporting documentation being impeccable (including the quantum and the time/delay aspects), and

   (v) the Claim being able to withstand the deepest forensic scrutiny without falling apart in the slightest.

I submit that most Claims would be found wanting when tested against those key requirements. However, by the time most Claims are submitted, the Contractor is usually already very desperate to be granted at least the extension of time request. The handling of Claims should therefore be treated as a serious matter on a non-stop basis, not something that can be commenced at the drop of a hat. It must be considered that the Employer's Team will be ready to send a Claim flying back over the net to win the match if the Claim fails in respect of any of the preceding five essential ingredients.

---

4 Critchlow J., Farr A., Briggs S. et al. (2006). *The Great Delay Analysis Debate*. The Society of Construction Law.

5 Lavers A. (2006). *The Great Delay Analysis Debate*, Epilogue, p. 52. The Society of Construction Law.

Chapter 9

Identifying Hazards and Managing the Risks

9.1   Introduction

The purpose of this chapter is to provide some background information, without going into too much detail, as to how the specific risks attaching to a Project can be both identified and managed properly. Risks arise as a result of hazards being present. A hazard has first to be identified, after which the level of risk associated with it then has to be assessed. Each area/aspect of a Project should therefore be analysed to identify and assess the major hazards and their associated risks. Any hazard posing a major risk must not be ignored. Having identified the primary risks involved with the major hazards, appropriate mitigation, control and monitoring plans then need to be put in place to deal with those risks. Minor hazards are those where minimising the risks can usually be handled effectively by applying common sense.

Nowadays, specialist Health, Safety, and Environmental (HSE) teams will handle the risks associated with the enormous amount of physical work activities that occur with every Engineering, Procurement, and Construction (EPC) Project. The hazard identification (HAZID) for the majority of such risks would have been completed a long time ago within most companies, and the mitigation measures required are therefore already very well understood and documented. This means that those outside the HSE teams only have to follow the safety rules that are already pre-set, and which the HSE teams will reinforce with appropriate Toolbox Talks/Training sessions. Nonetheless, for the benefit of everybody I consider that it is a good idea for all Managers on a Project to have a reasonably sound idea as to what goes into identifying hazards and managing their associated risks.

Generally therefore, with the exception of where physical work is involved, the majority of the individual Departments do not face significant hazards/risks that would impact their own internal work; their risks are predominantly negligible. However, where third parties are responsible for major sections of the implementation work (such as external Engineering Subcontractors responsible for the bulk of the design work, or Vendors manufacturing key items of specialised equipment), those items certainly need to be monitored closely. This is because they run the risk of adversely impacting the activities of other Departments too if things go wrong. Such items of concern should therefore be elevated to what is known as the 'Project Main Risks Register', so that they will then be completely visible to and handled by the Contractor's Project Management Team. It would be unusual for even a mega Project to have more than 50 major risk items that needed to be specially monitored, unless there were a number of different Site locations involved.

To decide what items should make it to the Project Main Risks Register, it is necessary to carry out a formalised and rigorous HAZID and risk management assessment exercise. The main benefits of doing this are that it will:

(i) provide information and data that will help improve the decision making about the perceived major risks, as well as create an understanding of the relationship between the components at risk (costs, time-frames, quality, safety, and the environment) and bring realism into the consideration of the trade-offs between them;

(ii) make the major risks, and the necessary actions to be taken to handle them effectively, clearly visible – not just to the Contractor's Project Management Team but to all other Managers too, including the Contractor's Corporate Management Team;

(iii) enable early solutions to be found for reducing both the likelihood of a risk materialising and the extent of its negative impact if it does materialise;

(iv) encourage the proper handling of risks, and thereby avoid having to manage a later crisis instead;

(v) ensure 'ownership' of risks, so that they will be more effectively monitored and managed;

(vi) help to improve the quality and accuracy of cost estimates and the Project Schedule; and

(vii) optimise the discovery and exploitation of opportunities at an early stage.

The effectiveness of all mitigation measures adopted must be tracked on a set, regular basis, because most mitigation measures require the input of others to make them effective. If the mitigation measures for dealing with a major risk are found to be ineffective only at some much later date, then the chances of implementing a successful catch-up plan will be greatly reduced. Monthly checking is usually too long a gap to measure the effectiveness of mitigation efforts, and weekly checks are much the best for catching problems before they have a chance to do irreparable damage.

9.2 Potential Hazards for Construction Projects

The major hazards that could impact construction Projects exist in a number of different forms, and the following is just a small example of where they sit:

1. **Delay Hazards**

Some hazards that have the ability to stop the Project's progress are not visibly evident but nonetheless exist, and may cause the Contractor severe problems. These are such issues as delays in engineering design, delays in procurement of key equipment, goods, and other materials, as well as construction delays, commissioning delays, and start-up delays. Most of those potential delays can be controlled to some extent or other by the Contractor. Other causes of significant delay may be very visible once they occur but are completely uncontrollable, such as flooding or sustained inclement weather conditions.

2. **Security Hazards**

   In countries where the political landscape is not stable, there may be the potential for an armed attack on the camp or the worksite, thereby putting the security of the personnel at grave risk. A certain degree of prevention measures can be put in place, but there will be limits on the mitigation level that can be achieved at a reasonable cost.

3. **Safety Hazards**

   There will be everyday hazards that impact the safety of the on-Site construction workers, such as working at height, operating heavy equipment, lifting heavy loads, working in open trenches or in confined locations, pressure testing pipework and equipment, energising high voltage electrical equipment, etc. All of these hazards can be contained by implementing safe working practices, providing adequate training and being vigilant.

4. **Environmental Hazards**

   There could be hazards that will impact the local environment, such as spillages of toxic materials, chemical leakage into water systems, etc. It should be possible to contain these hazards too by implementing safe working practices, again combined with adequate training and vigilance. However, there is a greater risk that something untoward might happen, primarily because the vigilance aspect may be diminished due to the activities involved taking place over a long time-frame.

5. **Quality Hazards**

   Some quality hazards can have major impacts. For example, if the wrong Welding Procedure Specification is applied to a critical area of work, then that work will need to be redone before any commissioning activities can be commenced. If it applies to work that has been buried (and perhaps already covered with a berm), the resultant lost time could prove disastrous for the Contractor. The Quality Control personnel must therefore be particularly vigilant, not just about the physical work being done but also to ensure that the correct specifications and work processes/procedures are being applied.

The following is an example of the major risks that could affect the well-being of a typical Project (as distinct from physical risks that could affect the well-being of participating personnel):

- Delayed Engineering Outputs;
- Late Placement of Purchase Orders for Long-Lead Items (LLIs);
- Manufacturing Delays for LLIs;
- Transportation Problems and Late Delivery of LLIs;

- Importation Problems for LLIs;
- Key Equipment Installation Problems;
- Equipment Interface Problems;
- Punchlist Close-Out Management Inadequacies;
- Connection Problems to Utility Suppliers; and
- Facility Start-Up Problems.

The hazard situation will also be subject to change over time, and items that were previously discounted as being of no significance may become elevated to a critical status, while items viewed as being potentially dangerous to people or the Project may not have materialised into actual problems. For this reason it is essential that personnel are designated as part of a team to be responsible for continuously monitoring the changing risk situation, re-assessing the risks as the Project progresses and putting in place mechanisms for dealing with the hazards and mitigating the risks wherever possible. This work should be conducted under the leadership of the Risk Manager. Where a Project is not of sufficient size to warrant the appointment of a full-time specialist Risk Manager, the overall responsibility for risk management should fall to the Project Manager. However, for the purposes of this chapter, reference hereafter will be made only to the Risk Manager as the key management authority for handling a Project's risk management activities.

As touched on above, the majority of hazards applicable to construction work are what can be described as 'commonplace' hazard situations, meaning that they are everyday realities, very often linked to physical work activities. The risks associated with such hazards are generally well documented nowadays and taken care of by the HSE personnel under the responsibility of the HSE Manager. The best way to mitigate the adverse effects of such hazards is to:

(i) ensure that the workforce is composed of competent people;

(ii) undertake regular, meaningful, on-the-job 'Toolbox Talks' and specific training to reinforce the need to be constantly vigilant against such hazards turning into accidents that could have been avoided;

(iii) provide sufficient numbers of full-time, dedicated HSE personnel on the worksite (at least 1 for every 50 workers), to ensure that there is effective policing of the work activities to counter the complacency that often sets in with repetitive work routines; and

(iv) ensure that all safety precautions that need to be taken (as are specified in the Method Statements, etc.) are fully implemented, and that all required safety equipment and protective gear is being used properly.

9.3 Responsibility for Project Risk Assessment

Ideally, a Project Risk Assessment should have been carried out by a company's top management personnel prior to the submission of the commercial bid. Once a Project has been secured, such an initial list of major risks for the Project should then be reviewed immediately from the very outset of the Project. The objective is to compile an updated list of the major hazards/risks perceived and fix the methods to be adopted to manage those risks. Such things to be considered are the expected security situation at the time the on-Site work is likely to commence and whether the local labour force being relied upon will still be available when required. All additional major risks identified (i.e. any that could threaten the successful achievement of the Project's objectives) should be brought to the attention of the Risk Manager for inclusion in the Project Main Risks Register.

To enable the Risk Manager to be as effective as possible in regard to risk management, a Risk Assessment Team should be set up, headed up by the Risk Manager. As a minimum, the Risk Assessment Team should comprise the Project Controls Manager, the Engineering Manager, the Procurement Manager, the HSE Manager and the Project Contract Manager (or, if one is not appointed, the Contract Administration Manager). As the Project progresses, other essential personnel should be added into the Risk Assessment Team as soon as they have been appointed, such as the Construction Manager, the Sub-contracts Manager and the Commissioning Manager.

Generally, before the commencement of a Project, the execution strategy and the schedule requirements (including an appropriate Work Breakdown Structure) will have already been established in sufficient detail to enable meaningful risk assessments to be made. Nonetheless, each individual Manager should be charged with the responsibility for considering all aspects of the work activities under his/her direct responsibility. They should also be made responsible for identifying the major risks that need to be entered into the Project Main Risks Register, both at the beginning of the Project and on a continuous updating basis as the work progresses. There will be many minor risks that the various Managers will be faced with as part of the everyday working of their Departments, and they are charged with finding appropriate ways to deal with those risks to ensure that they do not jeopardise the success of the Project. On the other hand, any major risks they identify that cannot be handled with the current level of resources in the Department must be elevated for inclusion in the Project Main Risks Register.

As each major risk is identified and its likely impact assessed, the Risk Manager must make a properly considered decision as to how much risk it is prudent to tolerate. The individual Managers must then strive not to exceed the agreed/accepted risk level in respect of each such risk, and report back immediately to the Risk Manager if things are not going according to plan.

There are two areas on a Project in which risks will be ever present on a daily basis once the physical work gets under way, those being (i) the health, safety, and security of the on-Site workers, and (ii) the quality of the work being done. To ensure that these two particular areas of risk are properly managed, two separate Departments are usually set up to deal with those specific objective, namely:

   (i) the HSE Department (to look after the safety, welfare, and security of the workers, as well as look after the environment), and

    (ii) the Quality Assurance/Quality Control (QA/QC) Department (to look after the quality of the design, procurement and physical work activities, as well as monitor the compliance level of the management teams and the workforce).

In respect of the safety and welfare of the workers, every reputable company must take this matter very seriously and strive never to compromise the safety and welfare of the workers in any way. To do this effectively means engaging adequate numbers of HSE personnel to micro-manage the safety aspects, guide the on-Site workers and monitor their activities. Guiding the workers will entail continually training and coaching the workers to be as safety conscious as possible. It will also require carrying out job risk assessments, 'Toolbox Talks' and training activities to equip the workers properly for the work that they will be required to perform.

All personnel working on a Project should be informed that they will be held responsible for taking steps to help eliminate all personal/personnel hazards as much as possible and, where hazards are inevitable, to help reduce the risks for all those personnel involved in hazardous working conditions. To that end, HSE Management Procedures and Security Plan documents must be compiled that set out the requirements for everybody's participation in the prevention of accidents and the protection of the welfare of personnel on the Project. The aim of the plan must be to reduce as much as possible the risk of injuries to personnel (and thus also avoid consequent lost time). Effective implementation and monitoring of such procedures and plans should be the responsibility of the HSE Manager.

Ultimately, however, the Project Manager has the responsibility for ensuring that all personnel on the Project fully understand the need to participate in risk avoidance and mitigation. The Project Manager must therefore arrange for the necessary personnel, facilities and other resources necessary to effectively implement, administer, monitor and enforce compliance with the various procedures that cover such issues, particularly those covering health, safety, the environment and security.

9.4   Identifying and Managing Project Risks

Central to the concept of effective Project Risk Management (as distinct from dealing with HSE hazards linked to manual tasks) is the selection of the most appropriate mitigation and control strategies to deal with the hazards observed. For each identified hazard it is necessary to take corresponding risk reduction actions in line with the severity of the risk level involved, in order to control it such that it does not exceed the agreed acceptable tolerance limit. Risk reduction solutions can include both managerial actions and technical actions, with the following being the generally preferred order of precedence for achieving risk reduction:

1. Eliminating the risk completely (such as by changing the sequence of work or employing an alternative design solution).

2. Transferring the risk to a third party (such as by taking out insurance or getting another party [specialists, perhaps] to assume the risk).

3. Putting appropriate mitigation plans in place to:

    (i) reduce the chances of the risk materialising, or

    (ii) contain the negative impact if the risk does materialise.

4. Accepting the risk 'as is', and then continuously monitoring it in order to be aware in good time if an alternative approach is becoming necessary.

If it is impossible to eliminate a risk (meaning that the hazard is impossible to avoid), then adequate contingency plans need to be put in place to limit the negative impacts and damaging effects of the risk in the event that it materialises. An example is the possibility of the worksite being struck by seasonal hurricanes/typhoons. The appropriate mitigation or control method is very context specific, and there is no universally right or wrong approach. The above generic considerations should be used for guidance but, ultimately, the decision as to the best approach to take will have to be based on the knowledge, experience and judgement of the members of the Project Management Team.

Full consideration must be given to all pertinent factors when the various Divisions and Departments of the organisation prepare the procedures and plans necessary for managing the work. Each Division/Department must be tasked with identifying the key risk issues applying to its area of work, and for producing its own unique set of procedures and plans to deal with those risks. Attachment A (Matrix of Project Risk Areas and Corresponding Risk Management Solutions) shows the principal risk areas that would apply to most on-shore construction Projects and the action needed to satisfactorily handle such risks.

As and when the heads of the various Divisions/Departments identify risks that they consider are exceptional and cannot be dealt with comfortably on a day-to-day basis, those risks must be raised up for the attention of the Risk Manager and included in the Project Main Risks Register. For all identified risks that are not elevated to the Project Main Risks Register, each head of a Division/Department is required to maintain a register of the risks specific to his/her own area of responsibility and to manage those risks on a continual day-to-day basis.

For most EPC Projects, there are in fact very few Departmental hazards that might turn into disasters and which therefore need careful consideration as to how best to handle them. The disasters that the Contractor must avoid at all costs are those that would destroy the Contractor's reputation. The Project Manager should insist that all these individual lists are prepared, because they may be later necessary to reduce the Contractor's liability and defend its reputation if an untoward incident occurs that results in injury or worse to workers.

The Departments that do run the risk of something very bad happening are the Construction Department and the Commissioning Department. Most other Departments face hazards that would have only limited negative impacts if the identified risks cannot be contained properly. The types of major risk faced by the Construction Department that are impossible to control run from severe inclement weather or situations caused by adverse weather conditions (such as floods, typhoons/hurricanes, tornadoes, avalanches, etc.) through to terrorist attacks on the workforce. Contractors in the Oil and Gas Industry face an additional problem of possible explosions occurring in situations where hydrocarbons are being introduced into the facility. Contractors specialising in mining and tunnelling work have to be careful that their operations do not have any negative impacts on the environment.

There are also many risks for which the Contractor is entirely responsible for controlling, where the level of control will depend to a great extent on the expected impacts if such risks materialise into actual problems. A decision therefore needs to be taken for each major risk to determine the most effective or appropriate actions to be taken for eliminating it altogether or reducing the severity of its impact so that it remains within a tolerable limit. Something along the lines of the Attachment B (Matrix of Prime Risk Considerations, Impacts, and Consequences) can help with that determination. The steps to take in the determination process are simple enough to list out (see below), but deciding upon the best course of action thereafter may not be so easy in many cases (and it may often be left for decision at a very high management level):

1. Assess the likelihood of the identified risk actually happening.

2. Determine the severity of the consequences that could arise if the risk materialises.

3. Decide what the most appropriate approach to handling the risk would be.

The severity level of risks and the control actions in respect of such levels of risk are generally classified along the following lines:

- *Minor risk.* An almost insignificant hazard that requires only a reminder about it before the activity commences.

- *Low risk.* A hazard that could most likely cause only limited disruption if it turned into an incident, and which needs only minimal attention to be paid in order to stop it developing into a problem.

- *Medium risk.* A hazard that requires constant attention to ensure that the standards and regulations common in the industry for the work involved are implemented properly at all times in order to avoid an incident occurring.

- *High risk.* A serious hazard that requires formulation of detailed control and reduction actions, and where an effective management programme needs to be in place in an effort to ensure that an incident does not occur.

- *Major risk.* A hazard that:

  (a) represents an intolerable risk (and which should therefore be considered as unacceptable), or

  (b) might cause considerable damage but be unavoidable and not controllable (due, for example, to external influences, such as a hurricane/typhoon), and for which emergency plans therefore need to be in place should an incident occur, or

  (c) could, in theory at least, be fully controlled but which would have disastrous consequences if the risk actually occurred due to failure of the risk control process.

9.5   Project Main Risks Register

A Project Main Risks Register should list out all the major hazards observed, along with specific details regarding the perceived source of the risks, explaining how the risks might materialise. The primary purpose of such a list is to allow a meaningful assessment to be made as to the significance of each risk in comparison with each of the other primary risks identified. It is essential that common day-to-day risks encountered in running a Project are kept out of the Project Main Risks Register. This is because there are myriad risks attached to undertaking even the simplest of Projects, but only limited time and resources available for an in-depth analysis of the risk issues by an organisation's top-level Managers. It is crucial that the major non-common risks are properly identified from the outset, so that adequate solutions for controlling the risks can be established. If such issues had not been picked up in the bidding stage, then adequate allowance for handling those risks might not be included in the Contract Price. That could mean that an innovative solution may be necessary to ensure that there are sufficient resources to handle a particular hazard that had been overlooked previously. However, the earlier the problem is identified the more chance there will be of the problem being handled satisfactorily.

The essential outputs required from working on the Project Main Risks Register are:

  (i)   the mitigation measures to be applied (or alternative approaches decided upon) to reduce the risks for each individual major hazard observed,

 (ii)   the identification of the ownership for handling each risk, and

(iii)   the allocation of the responsibility for monitoring the status of each risk and reporting back to the Risk Manager, in a timely fashion, in the event that the risk status begins to change from what had been anticipated.

Although ownership of the Project Main Risks Register lies with the Project Manager, maintenance of the register is the responsibility of the Risk Manager, who must ensure that:

  (i)   the register is continually updated in a timely fashion as the work progresses,

 (ii)   new risks are added as soon as they are identified, and

(iii)   previously existing risks are elevated, demoted, or closed out (as the case may be).

An indicative example of a Project Main Risks Register is shown in Attachment C (Example Project Main Risks Register), which was modified from the major risks identified during the bidding stage for an actual Project.[1] Once such a list has been compiled, it forms a good starting point for risk management after Project implementation has commenced, and it should be modified from time-to-time, as necessary, during the discussion sessions held by the Risk Assessment Team. The example shown indicates how the risk management process is documented, and comprises the following key elements:

---

1 Note: The list provided is a typical list only, not an exhaustive list. The risks for every project are different and will vary with a country's political climate, its economic status, its local geographical constraints, the transportation infrastructure serving it, etc. There may well be many more major risks on your new project than those shown in Attachment C; it needs very careful consideration.

(i) hazard observed (risk identification);

(ii) description of the proposed scenario and relevant context;

(iii) consequence scenario that will be used to assess the likelihood of the event occurring;

(iv) risk level assessment before mitigation (Relative Risk Factor);

(v) initially perceived impact level (H/M/L) on likely impact areas (cost/schedule/quality/Health, Safety, Security, and Environmental (HSSE)/reputation);

(vi) proposed mitigation measures;

(vii) residual risk level assessed once the mitigation measures are in place and working effectively;

(viii) impact level on likely impact areas anticipated after mitigation measures are in place;

(ix) identification of the Manager responsible for actively managing the risk and ensuring that the proposed mitigation measures are both in place and effective;

(x) identification of the risk owner who will be accountable if the risk is not effectively managed;

(xi) due date for the implementation of the mitigation measure;

(xii) action taken to comply with proposed mitigation measures;

(xiii) performance standard used to assess whether or not the mitigation measure is effective;

(xiv) evidence that the performance standard has been achieved; and

(xv) frequency of verification that the mitigation effort is still effective.

9.6   Risk Assessment Team Inputs

Once the Risk Manager has compiled the preliminary Project Main Risks Register, the Risk Assessment Team should convene a meeting to discuss each of the identified risks in detail, with the objectives of:

(i) properly defining each risk, so that its implications/ramifications are fully understood;

(ii) agreeing the assessed 'consequence' and 'likelihood' ratings for each risk, and ensuring that any proposed impact ratings are formulated correctly;

(iii) ranking those risks to establish the most significant (i.e. identifying those where it would be worth the team's effort to consider the matter more deeply, particularly if a positive benefit [opportunity] is present where an alternative design or working solution could be possible); and

(iv) agreeing on the control procedure and actions to monitor and/or mitigate each risk.

To ensure that a Project will be successful, priority should be given to the implementation and maintenance of risk mitigation measures for high consequence events and the Critical Path items identified in the Project Schedule. Further, the Project Main Risks Register should also be used to monitor the implementation of the actual risk control measures decided upon. If any high consequence mitigation measure is not working effectively (meaning that the risk is observed to be uncontrolled, with a high potential for negative impact on the Project), then the Risk Manager and the Project Manager must work together to establish an alternative and effective strategy as soon as possible.

To ensure that the true risk status of an item or activity is at all times properly assessed and understood, and that it is being responded to appropriately and positively, the Risk Assessment Team should meet at least on a once-a-month basis. The purpose of the meeting should be to review the Project Main Risks Register and to assess the effectiveness or otherwise of the mitigation measures that have been put in place. However, I have seen how easy it is, when the Project Implementation Team is under pressure, for the Risk Assessment Team's inputs to slacken off or even disappear altogether. This then led to new significant risks being overlooked until very late in the day, and which eventually required extra non-reimbursable funds becoming necessary to solve the problem. However, that situation could have been largely avoided if the potential problems had been brought up for discussion in good time.

Hazards and risks do not disappear because the Contractor is too busy to think about them – they seem to lurk around every corner, waiting to be overlooked before they finally heave into view, sometimes representing an unavoidable iceberg. This tends to happen far more where a Risk Manager is not appointed for a Project and, instead, risk management is left solely to the Project Manager to handle.

9.7 Relative Risk Factor Assessment

One method for assessing the relative ranking of the risks arising from hazards having varying degrees of negative impact is to allocate a number for each individual Hazard Effect (Consequence), ranging from 1 for the lowest impact to 100 for the highest impact. Using the following formula, combining the Hazard Effect with the probability of the hazard actually occurring, a Relative Risk Factor can then be calculated, as follows:

Relative Risk Factor = Hazard Effect × Probability

[where the maximum Relative Risk Factor would be 100%]

As an example, suppose we wish to assess the severity of the risk of Engineering delays impacting late delivery of Critical Path items. Based on past experience we may conclude that the probability of such a delay actually happening would be about 30%. We may also assess the Hazard Effect (the negative impact that would result if the potential hazard became a reality) to be about '80' (since the knock-on problems of having to pay Liquidated Damages for late completion would be severe). This would then result in a Relative Risk Factor of 24% (30% × 80).

By contrast, suppose we wish to assess the impact of unseasonal flooding. This time, we may consider that the Hazard Effect would probably only be around '40' (since, although the negative impact would primarily be to cause delay to the completion date, the 'Force Majeure' aspect of the event would most likely mean that Liquidated Damages for delayed completion would not apply). However, the chance of unseasonal flooding occurring (say once only in 30 years) means that the probability of such occurrence would be only about 3% if considering a 12-month construction period. Thus, the Relative Risk Factor would be only 1.2% ($40 \times 3\%$). The primary risk would be that the additional time (and associated costs) taken to recover from the effects of the rare flooding event would have to be borne by the Contractor.

Looking at the Relative Risk Factors of solely the above two examples, it can therefore be seen that much more effort should be concentrated on how to reduce the risks associated with Engineering delays than on how to mitigate the effects of unseasonal flooding. Of course, on a typical EPC Project there may be a large number of different major hazards to consider (perhaps up to 50 or so), but the principles of deciding on their relative seriousness and impact remain the same.

9.8 Risks Arising from Safety Studies

Some risks do not sit in plain sight but are hiding around corners, such as those that arise as a result of safety studies. Those studies are carried out in order to identify the risks to personnel who will later be operating equipment supplied and installed as part of the Permanent Works. For example, the following risk assessment activities, where needed, will be carried out by a team of people qualified in the specific area of work involved, all of whom will be erring on the safe side:

- HAZID (Hazard Identification);

- EHAZID (Electrical Hazard Identification);

- PHA (Process Hazard Analysis);

- HAZOP (Hazard and Operability);

- SIL (Safety Integrity Level); and

- ENVID (Environmental Impact Identification).

HAZID studies, EHAZID studies, and PHA are often chaired by a specialist independent (third party) person working in conjunction with the Contractor's Engineering Team and representatives of the Employer. Such studies aim to systematically examine and identify potential hazards associated with the proposed facility and recommend modifications to be incorporated into the design. The output of the HAZID and EHAZID studies should be recorded in a 'Hazard & Effects Register', which should be considered to be a live document that has to be continually updated during the subsequent stages of the Project. The Hazard and Effects Register should also form part of the 'Health, Safety & Environmental Impact Assessment Report' (HSEIA Report). All applicable recommendations in the reports should either be incorporated into the design or captured in the descriptions of the scope of work for the construction and commissioning phases.

A detailed HAZOP is normally conducted after the Piping and Instrumentation Diagrams (P&IDs) have been completed, just prior to freezing them. Usually, an independent HAZOP Chairperson will be appointed, with the Contractor's Discipline Engineers participating in the review in conjunction with representatives of the Employer. Sometimes the Contractor will be required to implement HAZOP recommendations arising at both the Front-End Engineering Design (FEED) stage as well as the Detailed Design stage (according to the stage at which the Contractor is appointed).

The Contractor will be responsible for tracking and closing-out all actions arising out of the HAZOP recommendations, and for issuing the HAZOP Close-Out Report well before the start of the construction work. Where necessary, interfacing of existing facilities with new facilities at different areas, and also interfacing between those areas, will need to be studied as well. Such interfaces must be identified at the beginning of the FEED (if the FEED work is the responsibility of the Contractor) and verified during the Detailed Design stage. An Interface Manager must therefore be appointed to ensure proper coordination between HAZOPs of different areas as well as between the Vendor Packages.

The Contractor must also obtain the P&IDs for all Vendor Packages, along with any related Hazard Study Reports prepared by the Vendors in advance of starting the HAZOP reviews of those packages. Any design changes following HAZOP action implementation and issuance of drawings previously approved for construction should be the subject of a further mini-HAZOP conducted in conjunction with the Employer's representatives. Failure to do this can lead to unexpectedly long delays in obtaining the Employer's approval of completed work.

After the HAZOP study and HAZOP recommendations have been incorporated into the P&IDs, an SIL determination study will often be conducted in accordance with IEC 61508.[2] This is done using the updated P&IDs, and the Contractor will be required to undertake verification of the assessed SIL study based on the Vendors' data. The Contractor will be required to implement all recommendations arising from the FEED and Detailed Design SIL studies, and also be responsible for the tracking and close-out of all actions arising out of the SIL study determination and verification recommendations. The 'SIL Study Close-out Report' should be issued well before the start of the construction work. In addition, the conducting of an 'Environmental Impact Identification Study' is sometimes required, although it may have to be conducted by third-party consultants approved by the Employer.

If all this work involved in safety studies and reports is not attended to closely, it all too often results in a great deal of time being lost in compiling completion readiness documentation. The problem often occurs because the Contractor does not take the requirements for such studies seriously enough, mistakenly believing that the Employer will be prepared to drop or simplify the requirements. This assumption could well result in unexpected delays for the Contractor that are impossible to mitigate, as well as

---

2 IEC 61508 (2010). *Functional Safety of Electrical/Electronic/Programmable Electronic Safety-Related Systems.* International Electrotechnical Commission.

add significant extra costs for engaging specialists who had not been considered by the Contractor at the bidding stage. In today's world, all such requirements for the studies stated in the Contract Documents must therefore be taken very seriously and properly incorporated into the Project Schedule. The time taken to conduct those studies, as well as the downtime for the Engineering Team while awaiting the outcome, can also be quite significant.

It is very important that the Contractor ensures that appropriately qualified personnel are appointed to represent the Contractor whenever any of the abovementioned studies are conducted. This is because, without doubt, the representatives from the Employer's side will be highly specialised and will push for the very highest level of safety to be incorporated into the Project, perhaps even where it is unnecessary. This can come in the form of such things as the Employer's Team insisting upon additional explosion proof electrical accessories being added. Another example is where additional expensive control valves are demanded that will take time to manufacture. If they are also required to be provided with signal links to the computerised control panels, that will add even more costs. The list of extras that could be requested is endless. Unless the Contractor has highly qualified personnel at the meeting who are knowledgeable enough to successfully challenge the reasoning put forward for such unwarranted extras, the Contractor will inevitably end up spending a lot more than expected on non-reimbursable items that, although being 'nice to have', were never absolutely necessary.

9.9   Dealing with Safety Risks to On-Site Personnel

For Projects in countries where the political situation is unstable and perhaps volatile, by far the biggest safety concern for foreign personnel temporarily based on location will be the issue of personal security. The implementation of comprehensive security measures will therefore be essential, especially if construction camps are involved. This will most likely require the construction of physical barriers such as perimeter trenches, concrete breach-proof walls, security fencing, and watch towers, plus the provision of armed guards and armoured personnel carriers (the latter often being provided by a third-party Private Security Company).

Regardless of the prevailing security situation, there will always be many potential safety hazards facing personnel working on a construction Site. Such hazards, and the level of the risk attached thereto, will vary according to the type of work involved. To ensure that the risks to workers are reduced as much as possible, it is essential to faithfully follow a recognised approach to controlling the start and ongoing performance of work. To achieve the required degree of control of work, each and every team of workers must be led each day by a dedicated senior supervisory technician (Trade Foreman, Superintendent, Specialist Technician, etc.). That person's responsibility will be to ensure that the potential hazards involved in the activities planned for the day have been properly identified, and to arrange for the appropriate mitigating action to be implemented. The responsibility for applying for and obtaining the necessary Permit to Work (PTW) should also fall to the relevant supervisory technician (the Job Performer). The Job Performer should also be made the person responsible for conducting appropriate training to ensure that all steps in the PTW application process are fully understood and will be followed.

In order to identify risks to personnel undertaking construction work and installing equipment forming part of the Permanent Works, it is imperative that a Task Hazard Identification (THI) and, where necessary, a Task Risk Assessment (TRA) review are carried out. The team for conducting the THI should ideally comprise the Construction Manager, the relevant Discipline Superintendents (specific to the type of work involved) and the HSE Engineer responsible for monitoring the particular job to be undertaken, all of whom should be qualified by experience in each specific area of work involved. In identifying the necessary controls to be implemented to contain the potential work hazards, consideration must be given to:

(i)   the nature of the job,

(ii) the Disciplines of the people involved,

(iii) the competency of the people undertaking the various activities involved,

(iv) the tools, construction equipment, and materials to be used, and

(v)   the working environment (including working height, confined spaces, presence of water or compressed air, potential for dangerous gases, etc.).

The steps to be taken when conducting a TRA are shown in Attachment D (Steps in Task Risk Assessment). The TRA team should attempt to identify the potential negative consequences, and list all the possible causes in a register similar to that shown in Attachment E (Task Hazard Assessment Worksheet). For each possible cause of risk, a corresponding control measure must be identified, specifying when and by whom the action ought to be taken (shown in the adjacent columns in Attachment E). Such reviews need to be undertaken (i) before carrying out any new job; (ii) if there is a significant change in the environment or conditions under which any previously assessed work is to be carried out; or (iii) where any work requires a PTW to be issued before the work will be allowed to commence. Having thoroughly assessed the hazards and risks involved, a decision must then be made as to the acceptability of those risks. If the risks are patently too high, then an alternative approach needs to be found for carrying out the affected activities. Even for activities with minor risks, mitigation measures should be put in place in an effort to ensure that those risks never materialise.

The TRA team must take into account not only how the work is to be conducted in the normal state but also what could occur in an abnormal or an emergency state. Based on investigation/analysis of (i) the circumstances under which the on-Site work is going to be performed and (ii) the records of previously collected data about similar work activities, the anticipated performance of the workers must be properly considered. This is in order to aid identification of as many as possible of the practical and the potential risk factors for the on-Site workers, such as:

1. **Unsafe physical conditions**, including the lack of (or defects in) such items as the protection to be afforded to the workers, the essential signalling or communication arrangements, the construction equipment, tools and accessories, the Personal Protective Equipment, and the quality of the environment in the specific work area.

2. **Unsafe human behaviour**, including the failure of the safety devices caused by operational error, the use of unsafe equipment, operation by hand instead of tools, improper storage of the object (referring to the finished product, semi-finished product, materials, tools, etc.), and venturing into dangerous places, etc.

3. **Unsafe working environment**, having the potential to result in personal injury, occupational disease, poisoning, and so on, including the physical factors (such as noise, vibration, humidity, radiation, etc.), chemical factors (inflammable, explosive, toxic, hazardous gases and oxides, etc.), as well as biological factors.

4. **Inadequate safety management**, including safety monitoring, inspection, accident prevention, emergency management, inappropriate job placement, and management of processes and operating methods.

5. **Mismanagement of discharges**, such as pollutants and solid wastes being released into any body of water, the atmosphere or the land, and energy releases (such as heat, radiation, noise, and vibration).

6. **Environmental problems**, including any manufacturing processes, packaging and transportation, waste management, acquisition and distribution of raw materials and natural resources, distribution, use and dumping of products, adverse impacts on wildlife and biological diversity, etc.

7. **Social mismanagement**, which includes both the failure to involve the local communities in the work associated with the Project and the failure to recognise the impacts on the local communities (such as not taking account of local customs, holidays, etc.).

The TRA team should consider the controls that it would be essential to impose for each specific activity to ensure that the work can be undertaken safely, and then document such requirements for handing over to the Job Performer. It must be remembered that job safety analysis is meant to have a practical application, not be relegated to a desktop and recording exercise only. Particular emphasis therefore ought to be placed on physical or procedural controls that prevent or reduce risk, for which contingency/emergency arrangements must be put in place to control or mitigate the consequences if anything does go wrong.

When all the controls for a task have been identified, the TRA should be redone to establish the level of residual risk expected to still remain once all those controls have been put in place. In particular, once all the control measures to reduce the risk have been identified, the following questions should be asked:

1. Have all the necessary control measures been fully/effectively identified?

2. Are any additional competencies required to complete the job?

3. Will the risk be controlled effectively?

If the level of residual risk for a particular task is considered to be unacceptable, consideration of the mitigation of the risk must be undertaken again for that activity. Once the process has been completed to reduce the risk to an acceptable level (ALARP – as low as reasonably practicable), the findings should be recorded and the new residual risk rating noted. Reducing the risk to the level of ALARP is the fundamental principle of risk management. However, if further control measures cannot reduce the risk to an acceptable level, the task must not be allowed to proceed, and the matter must be referred to the relevant Manager for a decision as to what steps should next be taken to resolve the situation. It is essential that agreement on risks and controls/mitigations is unanimous within the whole team. Once that has been decided, the TRA team must then proceed with documenting/recording those agreements and presenting them to the relevant Manager for review and approval. Once approval has been obtained, the information must be issued to the concerned Departments to follow/implement.

Key Performance Indicators (KPIs), a Project Improvement Plan and a Meeting Schedule should be developed within the Risk Management Plan. If poor KPIs indicate that a particular THI or TRA has probably been inadequately or incorrectly assessed, then that THI or TRA ought to be redone with the aim of finding a better solution for performing the task. Regardless of the relevant KPIs relating to task performance, if the risk control/mitigation for a particular task is subsequently perceived to be inadequate then the task must be stopped. Better risk control/mitigation measures must thereafter be both established and implemented (for example, the need to install mechanical ventilation and/or provide more lights).

Sadly, construction workers need to be constantly and frequently reminded of the dangers that beset them, no less when they are carrying out tasks with which they are very familiar, since complacency can soon set in. Consequently, it is essential that 'Toolbox Talks' are held before starting the work every morning, which should be conducted by the foreman (or anybody else appointed to be the Job Performer) with all the workers in the team present. The special hazards and control measures must be fully discussed to make sure that all attendees understand the nature of the work, the risks involved and how to reduce those risks.

For most companies today, there will usually be in-house soft copy files readily available that contain all the risk assessment details for almost all of the standard work tasks, so that the above steps do not have to be repeated for the bulk of the tasks on a new Project. The relevant risk assessment data (including the mitigation measures to be employed) can simply be printed out and compiled into a separate document for use on the new Project. However, when new, unusual and/or out-of-the ordinary tasks arise, then undertaking such risk assessments will be essential. An example would be where a river wall needs to be replaced completely along the frontage of a major river, where spring tides regularly take the water level almost to the full height of the wall. Breaching the wall under that particular scenario could easily turn into a complete disaster unless a separate temporary barrier of some sort is put in place to hold back the water. That was the dilemma faced by the designers when the New Hibernia Wharf building was constructed in front of Southwark Cathedral on the south bank of the Thames in London (completed in 1983).[3]

---

3 Now called Minerva House.

9.10 Dealing with Health Matters for On-Site Personnel

Major Employer bodies (especially the International Oil Companies) will generally have a suite of Management Procedures containing a comprehensive set of practices and requirements, all of which are intended to guide Contractors towards implementing safe working practices and ensuring good quality HSE performance. The Contractor will generally be required to prepare any necessary bridging documents to fill in the gaps between its own HSE Management System and the Employer's Management Procedures.

For a Project to be successful, it needs the input from a large group of healthy workers. The workforce for most overseas Projects will comprise many incoming foreign workers supported by large teams of local workers. The Contractor should take all necessary steps to ensure, before their arrival at the Site, that all such workers are sufficiently fit enough to be able to undertake the work that will be required of them. This is primarily because the medical facilities currently available in many developing countries are not to a high standard. The Contractor should therefore insist that all foreign workers have to undergo a thorough medical examination in their respective home countries before they depart for the Site, and their medical reports should be thoroughly reviewed by the Contractor's experienced HSE personnel.

In order to help in reducing the health risks to the workers, it would be prudent for the Contractor to provide access for all workers to the following medical facilities at all worksites:

(i) first aid kits and other basic medical equipment;

(ii) a suitable clinic or medical room for the temporary housing of sick workers;

(iii) a competent medical team capable of providing a minimum of basic First Aid Level-1 treatment;

(iv) advanced First Aid Level-2 response facilities available within 10 minutes (or, where the worksite is more than 60 minutes from a local hospital, a paramedic [or equivalent] ought to be provided full-time on the Site); and

(v) access to good medical care at a local hospital within 90 minutes of an emergency medical incident occurring.

The Contractor should also provide all its own expatriate employees with a valid medical evacuation scheme to facilitate timely departure to a nearby country with suitable medical facilities, and on to the individual's country of origin (should such a need later arise). The Contractor should also ensure that all its Subcontractors likewise have appropriate arrangements in place for their own expatriate employees. Further, to ensure that the risks to workers are reduced as much as possible, the Contractor should also put in place, and rigorously implement, an Alcohol and Illegal Substances control policy, and ensure that random checks are regularly made on the actual compliance level.

9.11   Dealing with Risks to the Environment

Employers will very often be required to sign up to environmental obligations with Government Agencies, in order to be granted permission to carry out high-risk operations in overseas developing countries. Protection of the environment and conservation of natural resources is of vital importance in such places, equally as much as in developed countries. The HSE Plans must therefore include provisions for making sure that the Employer does not breach any of its environmental commitments to any overseas government. Regardless of whether or not such government-imposed obligations exist, on any Project there is always the chance that the Contractor's activities could inadvertently cause harm to the local environment, which then badly impacts the local communities. To help reduce the risks involved, the Contractor must develop and implement environmental and social plans/procedures to deal with such issues, and then also integrate them into the HSE plans.

However, preparing environmental and social plans is only the first step to ensuring that the local environment and local communities will be well protected while the construction work is ongoing. What ideally needs to be done is for the Contractor to engage somebody full-time to monitor that all the required implementation steps are in fact being taken, in order to ensure that the local environment and communities will be protected properly. Environmental upsets can occur in various ways where construction work is taking place, especially if the work is associated with oil and gas pipelines. This is because spillages of any sort could pollute local water supplies and destroy vegetation in the wild as well as agricultural crops. It can even kill livestock. It is therefore important that, at the very least, the Method Statements for all work that affects local soil, vegetation or waterways should be reviewed, to see if there are any areas that could go badly wrong and lead to an environmental pollution problem or social problem occurring.

It is also imperative that, whenever work near live oil or gas pipelines is taking place, every precaution is taken to ensure that accidental rupturing of live valves does not occur. On one project I was working on, a mechanical excavator operated by the Employer's team was allowed to work close to the existing co-mingling oil pipelines, which then clipped a 2" valve. Oil containing significant amounts of hydrogen sulphide shot 300 m into the air and drenched the Contractor's workers who were working just 30 m away, causing many of the workers breathing problems. Fortunately, the evacuation procedure was sound, and all workers escaped without major injuries or health problems occurring. Since the accident was in a remote desert location, the damage to the environment was contained and did not affect any local communities or water supplies. However, using mechanical equipment so close to a live oil line was bordering on recklessness. Had that incident occurred in a pumping station close to a local community, the result could well have been disastrous for the villagers, as well as for both the Contractor and the Employer.

On another occasion, when I was working as part of the team for the Project Management Consultant (PMC), one of the Contractors working on a government-backed

Project needed some additional temporary space to store pipeline materials in a remote area. Another of our PMC team members authorised the cutting down of some 200 trees to create the required space, the Contractor having supposedly conducted due diligence checks to make sure that it was government-owned land. It was very fortunate that our local Environmental Specialist got wind of the intended action and intervened to stop the work before it had gone too far. The land was not government-owned as had been thought but, instead, was privately owned by people in the nearby village. The Contractor only had to replace 20 trees but was also required to pay compensation; it could have been much worse.

The lesson to be learnt from this is that an overseas Contractor can easily fall into the trap of making a mistake that will negatively impact the environment or the local communities. Often, this will be due to a lack of appreciation of the locality where the work is being performed and the communities who live there. This situation is often made more difficult because of the different language used in the area than the Contractor's team speaks. It would therefore be prudent for a Contractor to engage a local Environmental Specialist as a means of reducing the chances that environmental problems will arise. That same specialist should also be required to become involved in and comment on any Method Statements that are prepared for construction work involving disturbing the ground, cutting down vegetation or impacting waterways. I suggest that the costs involved would be small compared with those required to remedy a major environmental problem caused by the Contractor not being aware of the risks involved.

Attachment A

Matrix of Project Risk Areas and Corresponding Risk Management Solutions

MATRIX OF PROJECT RISK AREAS & CORRESPONDING RISK MANAGEMENT SOLUTIONS (Sheet 1)

	1.0 Strategic Planning	2.0 Planning	3.0 Engineering Design	4.0 Procurement	5.0 Construction	6.0 Commissioning Testing & Start-Up	7.0 Operations & Maintenance
1. Corporate Environment	1.1 Determine Project Management Policy	2.1 Arrange Funding & Establish Approvals	3.1 Set Need for Design Entities (incl. subcontracts)	4.1 Determine Procurement Strategy	5.1 Approve Construction Strategy	6.1 Approve Commissioning & Start-Up Plan	7.1 Approve O&M Plan
2. Scope & Design	1.2 Establish Full Facilities Scope	2.2 Establish Work Breakdown Structure	3.2 Fix Designer Responsibilities	4.2 Determine Material Requisition Plan	5.2 Determine Subcontracting Plan	6.2 Set Test Packs & Loop Checking, Etc. Plan	7.2 Determine Scheduled Maintenance Activities
3. Change Control	1.3 Determine Procedures	2.3 Set Up Cost Control Team	3.3/4.3/5.3/6.3/7.3 Set Up Management of Change Procedures				
4. Time Management	1.4 Size Project Implementation Team	2.4 Set Up Planning Team	3.4 Set Durations for Design Entities	4.4 Set Programme for Procurement Activities	5.4 Set Programme for Construction Activities	6.4 Set Programme for Comm/Start-Up Activities	7.4 Set Programme for Scheduled Maintenance Activities
5. Quality & Inspection	1.5 Determine Specifications	2.5 Set Up QA/QC Team	3.5 Determine Inspection & Test Plan Requirements	4.5 Determine FAT & SAT Requirements	5.5 Fix QA/QC Team Requirements	6.5 Fix QA/QC Team Requirements	7.5 Fix QA/QC Team Requirements
6. Cost Management	1.6 Establish Team & Procedures	2.6 Set Up Project Controls Team	3.6 Monitor Design Process Costs	4.6 Monitor Procurement Process Costs	5.6 Monitor Construction Costs	6.6 Monitor Comm/Start-Up Costs	7.6 Monitor O&M Costs

MATRIX OF PROJECT RISK AREAS & CORRESPONDING RISK MANAGEMENT SOLUTIONS (Sheet 2)

	1.0 Strategic Planning	2.0 Planning	3.0 Engineering Design	4.0 Procurement	5.0 Construction	6.0 Commissioning Testing & Start-Up	7.0 Operations & Maintenance
7. HR Management	1.7 Establish Team & Procedures	2.7 Set Up HR Team	3.7 Appoint Lead Design Engineers	4.7 Appoint Procurement Team Engineers	5.7 Appoint Supervision Engineers	6.7 Appoint Commissioning Engineers	7.7 Appoint O&M Engineers
8. Communication & Reporting	1.8 Determine Reporting Procedures	2.8 Set Up Document Control Team	3.8 Set Up IT Facilities	4.8/5.8/6.8/7.8 Set Up Reporting Formats			
9. Design Safety	1.9 Set Design Safety Management Policy	2.9 Set Up Design Monitoring Team	3.9 Carry Out HAZID, EHAZID & HAZOP Studies	4.9 Carry Out HAZID, EHAZID & HAZOP Studies on Packages	5.9 Carry Out Constructability Reviews	6.9 Prepare Commissioning and Start-Up Plans	7.9 Prepare O&M and Scheduled Maintenance Plans
10. On-Site Safety & Insurances	1.10 Set HSSE Management Policy	2.10 Set Up HSSE Team	3.10 Set Up Monitoring & Control Systems incl. KPI's	4.10 Check Vendors' HSE Management Documents	5.10/6.10/7.10 Monitor KPI's and Prepare Method Statements & Task Risk Assessments		
11. Security	1.11 Set Security Management Policy	2.11 Establish Physical Requirements and Security Personnel Requirements	3.11 Prepare Drawings for Physical Requirements and SoW for Private Security Company	4.11 Prepare Requests for Quotations, Invitiations to Bid & Commercial Bid Evaluations	5.11/6.11/7.11 Construct & Maintain Physical Requirements Ensure Security Requirements and Instructions from Private Security Company Understood By All Personnel At All Times (with adequate training implemented)		

MATRIX OF PROJECT RISK AREAS & CORRESPONDING RISK MANAGEMENT SOLUTIONS (Sheet 3)

	1.0 Strategic Planning	2.0 Planning	3.0 Engineering Design	4.0 Procurement	5.0 Construction	6.0 Commissioning Testing & Start-Up	7.0 Operations & Maintenance
12. Purchases & Subcontracts	1.12 Determine Procurement & Subcontracting Procedures	2.12 Set Up Vendor & Subcontract Procurement Teams	3.12 Prepare Material Requisitions, Scopes of Work & Technical Bid Evaluations	4.12 Prepare Requests for Quotations, Invitations to Bid & Commercial Bid Evaluations	5.12 Organise Delivery Dates & Subcontract Construction Windows	6.12 Arrange for Technical Services Support from Vendors and Subcontractor Participation	7.12 Determine Spare Parts Requirements
13. Deliveries & Logistics	1.13 Set Expediting & Transportation Policy	2.13 Set Up Expediting and Logistics Team	3.13 Establish Delivery Date Requirements for All Long Lead items, Equipment & Materials	4.13 Set Up Receiving Centre, Arrange Transportation & Warehousing	5.13/6.13/7.13 Prepare Storage and Materials Handling & Movement Facilities		
14. Interfaces	1.14 Determine Interface Plan	2.14 Appoint Interface Manager	3.14 Establish Detailed Interface Requirements	4.14 Convey Interface Requirements to Vendors & Subcontractors	5.14 Set Up Interface Status Report	6.14 Ensure Timely Attendance of Interface Parties	7.14 Set Up Reporting Mechanisms for 3rd Party Interfaces
15. Special Issues Management	1.15 Set Management Policy	2.15 Set Up Special Management Teams	3.15/4.15/5.15/6.15/7.15 Arrange Special Teams As Required				

Attachment B

Matrix of Prime Risk Considerations, Impacts & Consequences

MATRIX OF PRIME RISK CONSIDERATIONS, IMPACTS & CONSEQUENCES

SPECIFIC CONSEQUENCES OF OCCURRENCE OF RISK

SEVERITY LEVEL OF IMPACT IF RISK OCCURS	Cost Impact [Loss]	Schedule Delay Impact	Quality & Technical Performance Negative Impact	Health, Safety & Security Negative Impact	Environmental Negative Impact	Reputation Negative Impact
MAJOR (Level 5)	> $10 M	> 9 Months Delay	Major degradation of system performance, such as Key Performance Tests failure, leading to rejection of completed facility, with extensive penalties payable.	Multiple fatalities (3+) or > 30 injuries requiring hospital treatment.	Major external incident resulting in long-term or permanent damage, or prosecution under Environmental Laws.	International and local media attention and outrage over incident, relationship damage with Owner and Company, or severe regulatory and/or legal enforcement or intervention.
HIGH (Level 4)	> $2 M-10 M	4–9 Months Delay	Significant degradation of system performance with high penalties payable.	1 or 2 Fatalities or >10 injuries requiring hospital treatment.	Significant external incident resulting in medium-term damage, or Stop/Prohibition Notice issued, or extensive remediation required with potential prosecution.	Significant interest group outrage over incident, short-term national media attention, relationship damage with some stakeholders, or significant regulatory and/or legal enforcement or intervention.
MEDIUM (Level 3)	> $150 k-$2 M	2–4 Months Delay	Moderate degradation of system performance, perhaps with medium penalties payable.	Permanent disability or several non-permanent injuries.	Moderate on-Site incident resulting in short-term damage and remediation, or Improvement Notice issued.	Prolonged regional media attention over incident, relationship damage with regional community, or moderate regulatory and/or legal enforcement and intervention.
LOW (Level 2)	$50 k-150 k	1–2 Months Delay	Limited problems with light penalties payable (if any).	Single or multiple recordable injuries or health effects.	Limited on-Site incident easily contained, or excessive flaring.	Short-term media attention over incident, relationship damage with local community, or limited regulatory and/or legal enforcement and intervention.
MINOR (Level 1)	< $50 k	< 1 Month Delay	Small-scale problems with only small penalties payable (if any).	Medical treatment, first aid, over-exposure, etc., but no continuing ill-health effects.	Minor on-Site incident easily rectified, or abnormal flaring.	Isolated and short-term disruption due to incident, or complaints from neighbours and local community.

Attachment C

Example Project Main Risks Register

EXAMPLE PROJECT MAIN RISKS REGISTER - (Sheet 1)

Updated on:

Ref. No.	Hazard Observed	Scenario or Context	Perceived Consequence	Relative Risk Factor before Mitigation	Initial Impact Level					Proposed Mitigation Measures	Residual Relative Risk Factor (after Mitigation)	Manager Responsible	Risk Owner	Mitigation Action Date	Mitigation Action In Place?	Revised Impact Level					Performance Standard to Monitor Mitigation Effectivity	Evidence to Show Performance Standard Achieved	Verification Frequency for Mitigation Effectivity
					Cost	Schedule	Quality	HSSE	Reputation							Cost	Schedule	Quality	HSSE	Reputation			
1.	Delayed Engineering Output	Engineering Subcontractor not performing	Significant Project Delay; large costs and LDs	24%	H	H	n/a	n/a	H	Make MDR the Deliverables List and link payment to actual progress	8%	Project Controls Manager	Engineering Manager	Contract Effective Date +10 days	Yes - Engineering Subcontractor working to MDR	M	M	n/a	n/a	M	Measurement of Output against MDR	Comparison of actual progress against programme expectations	Weekly measurement, with daily monitoring
2.	Late Placement of PO for LLIs	Agreed PO documents not ready in time	Significant Project Delay; large costs and LDs	15%	H	H	n/a	n/a	H	Engineering to give priority to MR preparation and Procurement to chase up early EOIs	5%	Procurement Manager	Project Manager	Contract Effective Date +10 days	Yes	L	L	n/a	n/a	L	Measurement of Output against MDR	Comparison of actual progress against programme expectations	Weekly measurement, with daily monitoring
3.	Late Placement of Subcontracts	Scopes of Work may be defined later than needed	Significant Project Delay; large costs and LDs	15%	H	H	n/a	n/a	H	One person must be assigned at Senior Management Level to coordinate all activities from EOIs to Award	5%	Construction Manager	Project Manager	Contract Effective Date + 120 days	In Process	M	M	n/a	n/a	M	Deliverables to be set against an agreed programme of work and monitored	Updated Register of Subcontract Procurement Status	Weekly measurement of progress once S/c Procurement Execution Plan in place
4.	Late Delivery of LLIs	Vendors not performing	Significant Project Delay; large costs and LDs	8%	M	M	n/a	n/a	M	Expeditors to visit LLI Vendors regularly	2%	Procurement Manager	Project Manager	TBA Later	TBA Later	L	L	n/a	n/a	L	Receipt of regular Reports from Expeditors	Comparison of actual progress against programme expectations	Monthly by Expeditors
5.	Problems with Importation of LLIs	Documentation incorrect or lacking	Significant Project Delay; large costs and LDs	5%	M	M	n/a	n/a	M	Engage Import experts to guide and monitor activities	2%	Procurement Manager	Project Manager	TBA Later	TBA Later	L	L	n/a	n/a	L	Expected clearance times will be set and closely monitored	Comparison of actual clearance times against expected times	Daily by on-Site Procurement team after Goods arrival in country
6.	Equipment Interface Problems	Essential equipment is missing	Small Project Delays; incurring additional costs and possibly LDs	5%	M	M	n/a	n/a	M	Appoint Interface Manager to set a programme for and chase up all major interface activities	1%	Interface Manager	Project Manager	Contract Effective Date +10 days	Yes - Interface Manager appointed	L	L	n/a	n/a	L	Receipt of regular reports from the Interface Manager to update status of all major interfaces	Expected progress on interface fronts is proceeding as planned	Form time-to-time after Contract start date but monthly once Detailed Design commences

EXAMPLE PROJECT MAIN RISKS REGISTER - (Sheet 2)

Ref. No.	Hazard Observed	Scenario or Context	Perceived Consequence	Relative Risk Factor before Mitigation	Initial Impact Level Cost	Schedule	Quality	HSSE	Reputation	Proposed Mitigation Measures	Residual Relative Risk Factor (after Mitigation)	Manager Responsible	Risk Owner	Mitigation Action Date	Mitigation Action In Place?	Revised Impact Level Cost	Schedule	Quality	HSSE	Reputation	Performance Standard to Monitor Mitigation Effectivity	Evidence to Show Performance Standard Achieved	Verification Frequency for Mitigation Effectivity
7.	No-show by Equipment Supplier's on-Site Support Team	Security risks for personnel considered unacceptable	Small Project Delays; small costs and possibly LDs	5%	M	M	n/a	n/a	M	Follow-up mobilisation plans closely plus have contingency plan in place for a local expert team to step in	1%	Site-based Procurement Team + Construction & Commissioning Managers	Construction Manager	TBA Later	TBA Later	L	L	n/a	n/a	L	Receipt of timely status reports from Vendors about personnel mobilisation plans	Expected progress on personnel mobilisation front is proceeding as planned	Weekly across Vendors
8.	Punchlist Close-Out Management	Rectifying major defects is left until too late	Small Project Delays; small costs and possibly LDs	5%	M	M	n/a	n/a	M	Employ PMIS system from day 1 and appoint Punchlist Close-Out Coordinator	1%	C/A/QC Manager and Construction Manager	Construction Manager	TBA Later	TBA Later	L	L	n/a	n/a	L	Punchlist close-out plan for each Discipline to be prepared and updated in real time	Comparison of actual close-out progress against expectations	Weekly once population of Punchlist register commences
9.	Connection Delays for Electricity Supply at Substation	Equipment and 3rd Party personnel not available	Small Project Delays; small costs and possibly LDs	2%	L	L	n/a	n/a	L	Appoint Interface Manager to chase, all activities	1%	Interface Manager	Construction Manager	TBA Later	TBA Later	L	L	n/a	n/a	L	Receipt of specific reports from the Interface Manager on this critical activity	Expected progress on this critical interface is proceeding as planned	Weekly commencing at least 6 months before actual connection is due to be made
10.	Equipment Start-Up Problems	Inexperienced team takes too long to arrange and/or installation is incorrect	Small Project Delays; small costs and possibly LDs	2%	L	L	n/a	n/a	L	Engage Senior Equipment Expert to guide and monitor activities	1%	Procurement Manager	Commissioning Manager and Construction Manager	TBA Later	TBA Later	L	L	n/a	n/a	L	Regular updates required from Equipment Expert about mobilisation of personnel	Expected progress on personnel front is proceeding as planned	Weekly commencing at least 3 months before key personnel are required at Site
11.	Security Issues	Armed attacks directed at the Site and Camp and upon personnel in transit	Significant Project Delays incurring small costs (but no LDs)	5%	H	H	n/a	H	H	Engage Private Security Co., provide armoured vehicles + implement Owner's security recommendations + HSSE Manager will arrange training for entire workforce	5%	HSSE Manager	Construction Manager	ASAP for "Mobile" security (for Site Survey work) and Contract Effective Date + 150 days for "Static" security	Yes - "Mobile" security subcontract signed ("Static" security not yet required but negotiations ongoing)	H	H	n/a	n/a	n/a	Expectations for response times and manning requirements to be stipulated	Work alongside PSC and closely monitor response times and manning status	Ongoing on a daily basis following mobilisation of security teams

Updated on:

Attachment D

Steps in Task Risk Assessment

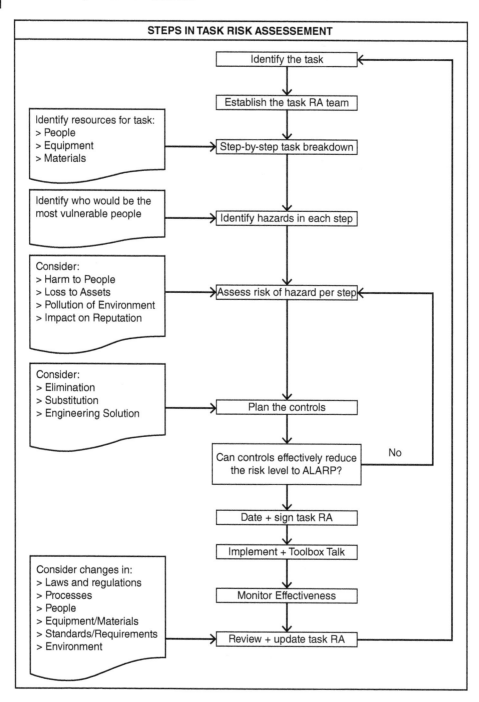

Attachment E

Task Hazard Assessment Worksheet

TASK HAZARD ASSESSMENT WORKSHEET

UNIT/PLANT:

SUBJECT:

DATE:

No.	RISK / ACTIVITY	POSSIBLE PROBLEMS + SCENARIOS	CONSEQUENCES	INITIAL RISK L	INITIAL RISK M	INITIAL RISK H	CORRECTIVE MEASURES/ACTIONS	WHEN	PERSON RESPONSIBLE	RESIDUAL RISK L	RESIDUAL RISK M	RESIDUAL RISK H	ALARP	ACTION?

RESPONSIBLE SUPERVISOR:

HSE ENGINEER:

Appendix A

Abbreviations and Acronyms

The following abbreviations and acronyms have been used in this book:

ALARP: As Low As Reasonably Practicable
BBSO: Behaviour-Based Safety Observation
BOD: Board of Directors
BOT: Build-Operate-Transfer
CAM: Contract Administration Manager
CAPEX: Capital Expenditure
CAT: Contract Administration Team
CLD: Corporate Legal Department
CMS: Completions Management System
EAMS: Enterprise Asset Management System
EDMS: Electronic Document Management System
EHAZID: Electrical Hazard Identification
EM: Engineering Manager
EOI: Expression of Interest
EPC: Engineering, Procurement and Construction
EPCC: Engineering, Procurement, Construction and Commissioning
EPCI: Engineering, Procurement, Construction and Installation
EPCM: Engineering, Procurement and Construction Management
ETN: Equipment Tag Numbers
EVMS: Earned Value Management System
FEED: Front-End Engineering Design
FIDIC: Fédération Internationale des Ingénieurs-Conseils (also known as The International Federation of Consulting Engineers)
HAZID: Hazard Identification
HAZOP: Hazard and Operability
HR: Human Resources
HSE: Health, Safety and Environmental
HSEIA: Health, Safety and Environmental Impact Assessment
HSSE: Health, Safety, Security and Environmental
IChemE: Institution of Chemical Engineers
ICMS: International Construction Measurement Standards
ICMSC: International Construction Measurement Standards Coalition

IEC:	International Electrotechnical Commission
IMR:	Interface Management Register
IOC:	International Oil Company
KPIs:	Key Performance Indicators
LDs:	Liquidated Damages
MDR:	Master Document Register
MOMs:	Minutes of Meetings
MR:	Materials Requisition
O&M:	Operation and Maintenance
OPEX:	Operational Expenditure
PCD:	Project Controls Department
PCM:	Project Controls Manager
PD:	Project Director
PEC:	Project Executive Committee
PEP:	Project Execution Plan
P&ID:	Piping and Instrumentation Diagram
PIM:	Project Information Manager
PM:	Project Manager
PMC:	Project Management Consultant
PMIS:	Project Management Information System
PTW:	Permit To Work
QA:	Quality Assurance
QC:	Quality Control
QD&E:	Qualifications, Deviations and Exceptions
SCADA:	Supervisory Control and Data Acquisition
SIMOPS:	Simultaneous Operations
SPIR:	Spare Parts Interchangeability Report
TBE:	Technical Bid Evaluation
THI:	Task Hazard Identification
TRA:	Task Risk Assessment
WBS:	Work Breakdown Structure

Appendix B

Glossary

In an attempt to make the technical content of this book as easy as possible to read, I chose to treat the words/terms listed hereunder as proper nouns. Other words/terms may also have been converted to proper nouns in the body of this book but which do not appear below, primarily because I considered that their meanings were perfectly clear as they stood. Some definitions I have provided will be quite obvious to the experienced practitioner, and I request such readers to bear with me. That is because this book is also aimed at younger, aspiring, construction-orientated managers and students of the construction industry, who wish to see how the members of Project teams interact and where activities occurring under EPC Projects can go wrong. I therefore trust that I will be forgiven for having included those more obvious definitions.

As-Built Drawing   The final version of a critical drawing (i.e. one that will be used after completion of the Project to aid in operating or maintaining the completed facility) following completion of the construction work, which has incorporated all the amendments to the original drawing as identified on the agreed/approved 'Red-Line' version of that drawing prepared in the field as the physical work progressed.

Base Camp   In overseas locations where a lot of individual worksites are anticipated/planned, it is usually a good idea to set up a centralised semi-permanent base to operate from. The type of facilities provided will usually include accommodation units and a canteen, warehousing, fabrication workshops, and even social activity areas. The personnel working in the offices during the daytime will generally include Procurement, Administration and Finance Department staff. Where there are armed conflicts in the wider area that would expose the workforce to possible attack, a Base Camp is essential. This would then provide a safe fall-back location if the personal security situation at the Site or at a Fly Camp looks as though it is likely to become dangerous.

Baseline Project Cost   The cost that the Contractor included in the Contract Price for undertaking and completing the whole of the work and services comprised in the Project. According to the accounting needs and principles within the Contractor's organisation, that may be a net or gross cost; whichever it is, it will be the cost that the Budget Controller must work to and not exceed for the Baseline Project Scope.

Baseline Project Schedule   The Contractor's initial programme and time-frame for undertaking and completing the implementation of the Baseline Project Scope, and which identifies all the key activities and the time-line allowed for each of those activities, together with the Critical Path(s).

Baseline Project Scope   The scope of the work and services defined in the Contract Documents at the time of signing the Contract that are necessary for undertaking and completing the implementation of the Project.

Bid Bulletin   A controlled document that is issued to all the bidders by the Employer to inform them of changes to the bidding documents or bid submission requirements.

Bid Enquiry Documents   Bidding documentation prepared by the Contractor and sent out to potential Subcontractors, which would generally define the scope of the proposed work and include drawings, specifications, Conditions of Contract, procedural/administration requirements for the Project's implementation, and forms/pricing schedules, etc. required for use when preparing and submitting Bid Proposals.

Bid Proposal   A formal submission sent in response to an invitation to submit a competitive quotation for supplying materials, goods and equipment and/or to carry out work or services related to a Project.

Build-Operate-Transfer   The term given to a business model where a concession agreement is entered into that grants an entity the right to occupy and operate a facility under prescribed conditions for a fixed period in exchange for designing and building the facility and assuming full responsibility for all the financing requirements up to the time of final handover. After the concession period has expired, the entire facility will usually have to be handed over to the Employer in its entirety, without any compensation being due.

Claim   A change that one party considers requires adjustment to be made to the Contract Price and/or the Project Schedule but which is not supported by the provisions of the Contract (neither wholly nor partly), the validity of which therefore needs to be established through discourse/debate between the parties in order to obtain agreement as to the cost and time impacts.

Commercial Bid Evaluation   An activity to establish whether or not the commercial bid submitted by a potential Vendor or potential Subcontractor is acceptable, and which is usually performed after the Technical Bid Proposal has been found satisfactory.

Commercial Bid Proposal   The priced proposal submitted by the Contractor in response to the Employer's issuance of an Invitation to Bid. It will usually contain the detailed prices comprised within the bid price and, quite often, is not required to be submitted until after the Contractor has been notified that its Technical Bid Proposal has been found to be acceptable.

Commissioning   The start-up of each of the individual pieces of the permanent equipment comprised in the Project after Mechanical Completion has been achieved.

Completions Management System   The generic term for a software system designed to keep track of all the documentation and data required by the Employer prior to handover taking place, the objective of that being to demonstrate that the technical integrity of the completed facility has been tested and verified all the way through the manufacturing and installation processes.

Conceptual Design The Employer's initial requirements for the completed Project (layouts, functional/operational requirements and the like documents), but which have not been developed to the stage where they can be considered to be Front-End Engineering Design documents.

Conceptual Design Validation Report A formal report submitted to the Employer that follows a thorough review by the Contractor's Project Implementation Team of the validity/suitability of the Employer's Conceptual Design documents. The purpose of such report is to identify/highlight possible anomalies, inconsistencies, inadequacies and omissions prior to commencement of the Front-End Engineering Design work.

Conditions of Contract Usually a document bearing that name, containing the topmost contractual terms and provisions applicable to the Contract, but which may in fact be lower in precedence order compared with the actual agreement document signed by both the Employer and the Contractor. Additionally, there may be other Contract Documents that inadvertently, or even deliberately, have an impact on the interpretation of the content of higher precedence Contract Documents (including the Conditions of Contract).

Construction All the effort/input and output in respect of the physical implementation work for building construction Projects, heavy or civil construction Projects and industrial construction Projects (e.g. for the oil, gas, power generation, petrochemical, pharmaceutical and manufacturing industries).

Construction Department The team that has the responsibility for undertaking all the physical construction work for a Project (both the temporary and permanent work).

Contract The formal signed agreement between the Employer and the Contractor in respect of the Contractor's undertaking to complete the Employer's Project.

Contract Documents The formal set of documents that collectively form the basis of the Contract between the Employer and the Contractor.

Contractor A business entity that earns its revenue by carrying out construction work.

Contractor's Team All the Contractor's management personnel engaged on a Project, from the Project Manager downwards (but excluding any third-party Engineering Subcontractor(s)).

Contract Price The amount of money that has been agreed at the signing date between the Employer and the Contractor as the Contractor's remuneration for completing the Project in full accordance with the terms and provisions of the Contract.

Contract Summary A 'quick reference' document, prepared specifically for the purpose of giving all the Contractor's Managers sufficient details of the fundamental elements of the Project and the Contractor's obligations under the Contract.

Corporate Management Committee A small group of senior people (usually comprising Directors and, occasionally, Senior Managers) set up by the Board of Directors for the purpose of vetting any major items that the Board of Directors considers should be reviewed independently from the teams actually working on Project tasks (such as, for example, Bid Proposal documents, large-value Purchase Orders or Variations, etc.).

Corporate Management Team   The Board of Directors and all the Managers and Sub-Managers of a construction company that are not directly engaged on Projects. (Very often, a Contractor will be led by a General Manager [or similar such person] who will be expected by the BOD to supervise and manage all the Corporate Managers, all of whom would be expected to report to the General Manager. In turn, the General Manager would then be the person responsible for conveying all important information/matters to the BOD. There will of course be variations of this set-up from one company to the next.)

Cost Management Team   The Cost Engineers and Quantity Surveyors allocated to the Project, who may be organised under the Finance Manager, the Project Controls Manager or even the Budget Controller, according to the Contractor's preferred management set-up (which may vary considerably, depending on the size and complexity of the Project).

Critical Path   A sequence of logically connected, interdependent activities that have no float within a networked project implementation programme, from the planned start date through to the anticipated finish date, and which results in the longest overall duration for achieving completion of the project.

Data Sheet   A document summarising the technical and materials characteristics of a product (especially for equipment, where the required performance outputs may also be stated), including connection details so that the product can be integrated with the other products comprised within the system or sub-system.

Deliverables   The drawings and documents that are specifically required by others, either for the purpose of reviewing/approving (such as by the Employer), or for passing onto others for inclusion in their drawings, or which are required for others to follow when carrying out work (such as for the Construction Team in the field). Such drawings and documents are 'must-haves', as distinct from being submittals that are provided for information only.

Department   A group of people operating within a specifically designated section of a Contractor's organisation under the direction of a Manager or Sub-Manager, responsible for carrying out a specified set of unique functions, either at the corporate level or the Project level.

Design Team   For non-EPC Projects, this will often comprise different entities such as Architects, Structural Engineers and Building Services Engineers to carry out the design work, construction monitoring and Project administration (usually under the lead of the Architect). For EPC Projects, it will usually involve a company with specialist Engineering design skills related to the type of facility required, which will undertake the conceptual design and Front-End Engineering Design (FEED). Sometimes, the Design Team responsible for the FEED may be engaged to administer the Project on behalf of the Employer, and is often referred to as the Project Management Consultant.

Detailed Design   The detailed drawings necessary for showing how the physical elements of the completed Project are required to be put together, which are developed from the approved/agreed Front-End Engineering Design documents. Those drawings are then passed onto Vendors for them to develop their own Shop Drawings (very detailed drawings required for manufacturing/production and installation purposes) and then, ultimately, to the Construction Team to follow for constructing the facility. The Detailed Design work also includes the preparation of all specifications for the physical elements of the completed Project.

Discipline   Any specialised branch of knowledge or type of work carried out within a Contractor's organisation that requires either study via a higher education establishment or appropriate specialised training received at technical schools/colleges and/or in the workplace.

Dispute   A Claim having a cost and/or time impact that is rejected by the receiving party because the other party believes that it does not appear to be specifically claimable under the provisions of the Contract, and which the originating party intends to continue pursuing until the matter is resolved, even if that means resorting to arbitration or litigation.

Document Control Team   The group of people in a company responsible for handling, storing, retrieving and disseminating incoming and outgoing documents in all their various forms (letters, formal emails, drawings, specifications, policies, procedures and the like). They will more than likely be aided by an Electronic Document Management System (EDMS), and will be responsible for interfacing where necessary with the EDMSs of others in order to pass on such documents to other organisations doing business with the company.

Earned Value Management System   A structured approach to measuring progress and cost performance of the Project at any point in time, and comparing that data against the original progress and cost/revenue expectations. The ultimate objective of its application is to enable a realistic forecast to be made of the Project's likely completion date and final costs. Its route to achieving this is by utilising the earned value of work done to measure progress objectively.[1]

Employer   The entity requiring a Project to be constructed, with the ultimate responsibility for specifying all the required objectives for the Project, and for appointing and paying the Contractor to undertake and complete the Project.

Employer's Team   The entirety of those representing the Employer, including the Employer's in-house representatives and any Project Management Consultants, Third-Party Inspectors, etc. appointed by the Employer.

Engineering Subcontractor   A third-party organisation undertaking any part of the engineering design work that the Contractor is obligated to produce under the Contract for the Project.

---

1 Based on both the *Earned Value Management Handbook and the Earned Value Management APM Guidelines*. Association for Project Management.

EPC Contractor   A business entity that specialises in undertaking construction Projects where design development is required. That may include preparing the Front-End Engineering Design (FEED) documents based on a Conceptual Design, preparing the Detailed Design documentation from the FEED documents or even completing partial FEED documentation prepared by others. It will also most certainly include: (i) preparing the Detailed Design documentation; (ii) procuring the bulk (if not all) of the materials, goods and equipment; (iii) undertaking the construction and installation work (including assuming responsibility for all specialist subcontracted elements); and (iv) either assisting with or directly undertaking the commissioning work before then handing over the completed facility to the end user or operator.

Float   A general term for the amount of time that an activity in a Project network can be delayed without causing a delay to either successor tasks (referred to as 'Free Float') or to the Project's contractually required completion date (referred to as 'Total Float').

Fly Camp   A temporary camp set up on or near the Site to house personnel and facilities that are to be used for the specific purpose of undertaking the construction work for a Project, and which will be dismantled immediately after the purpose for which it was set up has been achieved. In locations where a lot of individual worksites are anticipated/planned and armed conflicts in the wider area could expose the workforce to possible attack, it is also a good idea to establish a Base Camp in a relatively much safer location than where the worksites are located. This will then provide a fall-back position if the personal security situation at the Site and/or Fly Camp looks as though it is likely to become dangerous. A Fly Camp is essential if the drive back to relative safety in the dark after working hours would be particularly dangerous (where, sometimes, governmental restrictions may apply to travelling at night). This is especially so if it would be mandatory to employ an armed 'personnel security detail' to accompany every vehicle movement to and from the Site.

Free Float   The amount of time that an activity in a Project network can be delayed without impacting the start date of any of its successor activities.

Front-End Engineering Design   The documentation that sets out the Employer's full requirements for the layout, design, materials, goods and equipment regarding the construction of a Project's Permanent Works, and against which the Contractor is required to develop the Detailed Design work.

Functional Specification   Documentation that describes both the functional and operational requirements of a proposed facility, while at the same time generally avoiding specifying the specific materials, goods and equipment to be employed.

Guaranteed Performance Outputs   A statement of the output requirements for specific items of equipment and/or the completed facility that the Contractor must guarantee will be met or, in the event of failure to meet those requirements, the Contractor will be required to compensate the Employer commensurate with the loss of required output.

Hazard and Operability Study   A Detailed Design review technique employed by design and operations experts sitting together and analysing the design in detail, in an effort to identify and correct deficiencies in the design that may otherwise give rise later to safety or operability problems in the operational stage.

Information Handover   The entirety of the documentation that the Contractor will be required to hand over to the Employer, operator or end user to enable effective operation and maintenance of the completed facility, which information will usually be required to be handed over in full before the Employer will issue the Final Acceptance Certificate.

Information Management System   A software system designed to facilitate the storage, organisation and retrieval of information such that it can be accessed not only by people within an organisation but also by authorised people from other organisations. At its most basic level it can be used for tracking correspondence and for keeping control of Equipment Tag Numbers. For a Contractor involved in maintenance activities, it can extend to such things as warehouse management in order to keep track of the spare parts inventory and automatically re-order products as they are used up. For Employers, it can also be used to store details of all the equipment involved in the Project for the purpose of asset management.

Inspection Release Notification   A document that certifies that materials, goods or equipment that need to be shipped have been inspected and found to be in full conformance with the specification and other contractual requirements, and which the Employer usually insists must be issued prior to the Employer's Team authorising shipment.

Interface Management Register   A matrix identifying each individual Interface Issue – very often compiled in spreadsheet format – in which the key information for the purpose of tracking the processing and closing out of those Interface Issues is logged.

Interface Issue   A situation where an interconnection occurs between the work of the Contractor and an external third party, or between two or more Subcontractors/Vendors, and which needs to be resolved by raising information requests or obtaining clarifications/documents from another party, usually required by specific critical dates.

Job Performer   The person responsible for supervising a particular physical work activity, especially one where a Permit to Work is required prior to commencing the activity. The Job Performer will usually be required to be fully experienced in a specific Discipline allied to the work involved in that activity, be trained in the Permit to Work application procedures and be responsible for giving the appropriate Toolbox Talks/Training prior to that activity commencing.

Key Performance Indicator   A measurable (quantifiable) value used to establish the success (or otherwise) of an organisation and/or its employees in meeting important objectives for performance.

Lead Discipline Engineer   The person responsible for heading up a team of specialist design Engineers (such as for electrical, instrumentation, piping, civil work, etc.) and monitoring/guiding them to achieve the objectives set for the team.

Letter of Intent   A written interim agreement, summarising the main details for a proposed contract, usually stating that the formal contract will be signed at a later date. It is often issued to provide confidence that a deal is about to be struck. However, although it may not be intended to represent a formal contractual obligation, the language employed may inadvertently create a binding agreement that one or other party never intended, so great care must be taken in its drafting.

Liquidated Damages   The compensatory payment to be made to the Employer (and pre-agreed as being not a penalty but a genuine estimate of the loss that may be sustained by the Employer) in the event that:

(i) the Contractor fails to complete the Project by the contractual completion date (or any later extended date that may be applicable under the provisions of the Contract), and/or

(ii) the completed Project fails to achieve the Guaranteed Performance Outputs.

Local Content   An obligation to use and develop the skills of the local workforce and engage local companies as much as possible. This obligation is imposed by local governmental authorities on international Contractors wishing to be awarded work in a foreign country. Compliance with such obligation is being enforced far more rigorously nowadays, in order to avoid unrest from unemployed locals in the vicinity where a new facility is being built.

Long-Lead Item   Any item to be purchased that is expected to take a long time to procure (whether because of complicated design requirements or lengthy manufacturing times), or any work that will take a long time to construct and is on the Project's Critical Path because of the criticality of subsequent follow-on activities.

Master Document Register   The listing compiled by the Contractor to show all the drawings and documents that will be produced by the Contractor's Project Implementation Team, many of which (but not all) will be Deliverables required to be reviewed by the Employer's Team.

Materials Requisition   A description of the technical, functional and operational requirements of a product (including detailed specifications) that needs to be purchased for incorporation into the Permanent Works, and which will be issued to bidders as part of the Request for Quotation package.

Maturity Management System   An object centred means of progress measurement (as opposed to being document centred), enabling everybody to see the true status of each individual item of plant or equipment from initial identification, through tag numbering and up to and including testing and commissioning.

Maximo   IBM's Maximo 'Asset Management' software, which is an enterprise asset management product that some companies use to track the operation, maintenance and disposal of assets. It also allows operational data to be entered and stored, such as full details of all equipment (including the tag numbers and information related to replacement spare parts holdings, and also instructions for maintenance and servicing, etc.).

Mechanical Completion   The completion of all the pre-commissioning work following completion of the construction work that is necessary before the commissioning work can be commenced, especially before hydrocarbons can be introduced into the relevant process systems.

Method Statement   A document detailing the way that a work process is to be undertaken, outlining the hazards involved and providing a step-by-step guide as how the job will be undertaken safely.

Operations Team   The personnel who will be responsible for taking over the completed facility from the Contractor's Construction or Commissioning Teams and running/operating it.

Performance Bond   A document issued by a bank or insurance company that guarantees the beneficiary payment up to a specified total value if the beneficiary validly calls upon the bond at any time. It most certainly is not a guarantee that the Contractor will complete the Project satisfactorily (contrary to what may be found in Wikipedia[2]).

Permanent Works   All the physical property arising from the implementation and completion of the Project work.

Permit to Work   A system of work control to ensure that the commencement of hazardous on-Site physical activities does not begin until full approval has been given to the proposed implementation methods, including manpower and equipment resources, safety provisions, etc.

Personal Protective Equipment   Equipment worn to minimise exposure to hazards associated with certain physical construction work activities, such as but not limited to the following:

- gloves for handling rough products where hands could be damaged,
- steel-capped boots to protect feet from injury due to sharp objects being present at floor level,
- eye protection for welding and grinding activities,
- earplugs and the like to protect ears from damage due to loud or harsh noises (especially from running heavy plant),
- helmets to prevent damages from falling objects,
- breathing apparatus for when work takes place in areas where toxic fumes or products may be present, and
- full body suits to provide overall protection where toxic materials have to be used.

Plot Plan   A scale drawing showing the boundaries of a building site and the buildings, equipment layouts, utility services runs, roads layout and other construction work required for a proposed Project.

Project   A unique construction endeavour undertaken by a Contractor to provide a specific completed facility, having a fixed scope of work that is required to be completed within a set time-frame under a controlled budget, and which is aimed at achieving all the objectives prescribed by an Employer.

Project Execution Plan   A document that sets out the Contractor's plans for managing the Project from mobilisation right through to close-out of the Project. Its purpose is to enable people to understand quickly what the whole Project is about, as well as see

---

2 Wikipedia. *Performance Bond*. https://en.wikipedia.org/wiki/Performance_bond (accessed 28 August 2018).

what the specific roles and responsibilities of all the various participants involved in the Project are and how the specific work activities are to be implemented.

Project Executive Committee   A select group of Contractor's personnel appointed by the Board of Directors, charged with the responsibility for vetting the recommendations for the placement of all Purchase Orders and Subcontracts, in order to ensure as much financial control as possible while circumventing an unnecessarily long approval process.

Project Implementation Team   All the Contractor's personnel responsible for managing/administering the Project and participating in producing/preparing the Contractor's technical work for the Project (as distinct from those doing the actual physical work involved with the Permanent Works), inclusive of Sub-Managers, but excluding members of the Corporate Management Team.

Project Main Risks Register   A list setting down the primary risks that are particular to the Project, and which sit outside of the day-to-day risks that are common to most projects. The common risks are many, and encompass such matters as the safety of the construction workers, protection of the environment from the construction work activities, quality of the construction work, deterioration of public roads due to heavy construction vehicles, etc. Such common risks fall to full-time teams to handle, staffed by experienced and competent personnel, such as those in the Health, Safety and Environmental Department, and the Quality Control Department.

Project Management Consultant   An organisation that provides personnel to assist the Employer in some or all of the following:

- preparation of the necessary drawings and other documentation showing the basic requirements for the Project,
- organising and handling all activities involved in the EPC bidding process,
- monitoring the Contractor's implementation activities,
- reporting to the Employer on all aspects of the Project's implementation, and
- conveying the Employer's decisions to all other participants.

Project Management Team   The topmost level of the Contractor's management personnel charged with the responsibility to manage/direct all the activities necessary to undertake and complete a Project such that it satisfies all the objectives set by the Employer in regard to (i) the scope of work, (ii) the quality of the finished facility, (iii) the time-frame in which all those activities are to be undertaken, and (iv) adherence to the budget constraints and controls. In the process of managing/directing a Project, the Project Management Team is also responsible for satisfactorily identifying and managing all the risks involved, and for ensuring that the work of the Contractor is undertaken as safely and efficiently as possible, all while exercising good budget control.

Project Risk Management   The process of:

(1) proactively identifying the major potential hazards/risks for the Project as early on as possible (and revisiting/redoing that exercise regularly throughout the entire duration of the Project),

(2) properly analysing and assessing each of the identified hazards/risks in turn, so that sound commercial decisions can be reached as to what the most appropriate response to each should be,

(3) putting in place adequate mitigation measures where necessary to counter those hazards/risks, and

(4) regularly monitoring the effectiveness of employing the hazard/risk mitigation measures and making appropriate modifications if later those measures are found to be ineffective.

Project Schedule   The table-form document identifying all the Contractor's key activities/tasks and their associated time-frames to enable completion of the Project within the overall duration required by the Employer.

Project Scope   The totality of the contracted work to be undertaken in respect of the Project and which should have been fully allowed for in the original Contract Price.

Punch Items   Completed items of physical work that contain defects or are otherwise incomplete in some way and therefore need to be rectified/completed, as well as work items that still remain to be done in order to complete an item or element of physical work.

Purchase Order   The official order to the Vendor from the Contractor setting out the terms and conditions of the agreement between the Contractor and a Vendor in respect of the provision by the Vendor of materials, goods and/or equipment for a Project (and which may include on-Site support services in regard to installation work).

Qualifications, Deviations and Exceptions   A formal listing prepared by the Contractor that is usually submitted as part of the Technical Bid Proposal, setting down specific details for such things as:

(i) what the Contractor's bid pricing allows for in circumstances where the Contractor considers that the requirement was unclearly stated,

(ii) where the Contractor has priced for something different than was specified, and

(iii) what the Contractor has not priced for even though it may have been specifically required/specified in the Invitation to Bid documents.

Quality Assurance   The Quality Assurance (QA) function involves setting up the procedures/processes to control the quality of the outputs and the auditing of how the work was done, in order to check whether or not the required procedures/processes were followed properly.

Quality Control   The Quality Control (QC) function involves closely checking the quality of the Contractor's output to see that it matches the expectations of the established quality procedures/processes.

Red-Line Drawing   A construction drawing that shows corrections or changes that were found necessary in the field, and which will be used as the basis on which to prepare the As-Built version of the drawing.

Rely-upon Information   Information provided by the Employer that the Contractor can take as being correct (whether it is subsequently found to be the case or not) and upon which the Contractor's bid pricing and work programme can be deemed to have been based.

Request for Quotation   Bidding documentation prepared by the Contractor to send out to potential Vendors.

Service Providers   A business that provides support services to Contractors, such as road hauliers, Customs clearance services, etc.

Site   Any areas where activities associated with a Project's construction work take place, including the place where the Permanent Works are to be executed and also the associated Base Camp and Fly Camp, workshop and fabrication facilities/areas, maintenance and servicing areas, storage areas, laydown areas and the like.

SPIR   A form (the Spare Parts Interchangeability Report) that sets out all the details of the spare parts for the purpose of allowing the Operations Team to decide what spare parts to order for holding in stock (usually for the first two years of operation of the facility).

Start-Up   Commencement of the process facilities for which purpose the Project was constructed, after successful Commissioning of all the Project's systems and sub-systems.

Subcontractor   A business entity that has entered into a contractual arrangement for undertaking specific work activities or for providing services for the Contractor.

Task Scheduling   The iterative determination of the order/precedence in which all the separate tasks in the Work Breakdown structure need to be undertaken, and allocating the resources needed for each task and the time required to complete each task.

Technical Bid Evaluation   An activity to establish whether or not the technical bid submitted by a potential Vendor or potential Subcontractor is competent and acceptable, and which is usually performed prior to the opening (or consideration) of the corresponding Commercial Bid Evaluation.

Technical Bid Proposal   The unpriced proposal submitted by the Contractor in response to the Employer's issuance of an Invitation to Bid, containing details of the Contractor's technical capabilities and proposed technical solutions. It is quite often required to be submitted and judged to be acceptable before the Commercial Bid Proposal is required to be submitted.

Thin Client   A computer that lacks its own hard drive and must therefore run on application software resources stored on a remote server. This improves security, because such computers are restricted by the server as to what they can be used for. For example, they cannot run unauthorised software or save/copy data anywhere other than on the remote server, neither can they send any data to external locations.

Tier-1 Contractor   A construction company that is amongst the largest, wealthiest and most experienced in the industry, and thereby capable of successfully undertaking Projects worth billions of dollars.

Total Float   The total amount of time that an activity in a Project network can be delayed without causing a delay to the Project's contractually required completion date. It therefore represents the total amount of flexibility there is in the anticipated duration for completing an individual activity.

Variation   A change having a cost and/or a time impact that is specifically valid and claimable according to the provisions of the Contract, thus making it capable of being easily agreed between the Contractor and the Employer.

Vendor   An entity in the construction supply chain that provides goods (and, sometimes, installation and other services associated with those goods) that are required for a construction Project. A Vendor may not necessarily be the manufacturer of the goods themselves, but it is the party that is paid for the goods/services provided. Sometimes the term 'Supplier' is used in place of the word 'Vendor', and sometimes the Vendor may be an agent of the manufacturer.

Vendor Documents   All those essential documents that the Vendor is required to prepare and submit to the Contractor in order for the Contractor's obligations to the Employer to be fulfilled, including manufacturing drawings and documents, materials certificates, equipment test certificates, shipping documentation, SPIR Forms, etc.

Work Breakdown Structure   A comprehensive listing of all the activities and Deliverables that need to be accomplished/submitted in order to execute and complete a Project, set down in a logically structured and meaningful hierarchical manner, and subdivided into easily understandable and manageable sections of work.

In addition to the above, the following words/phrases used in the body of this book refer to publications, entities or universally-accepted/well-recognised terms that should need no further explanation or introduction:

- Bribery Act, 2010
- Foreign Corrupt Practices Act
- International Oil Companies
- International Organization for Standardization
- Oil and Gas Industry
- US Securities and Exchange Commission

Appendix C

EPC Project Management Team Organisation Structure

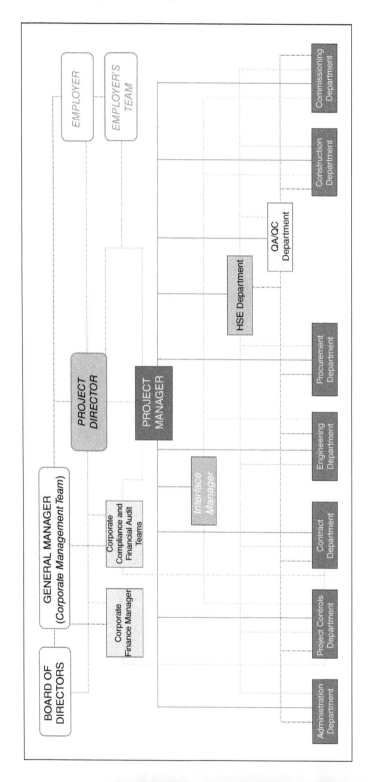

Appendix D

EPC Project Departmental Organisation Structure

Administration Department
Administration Manager
Administrator
Visa Coordinator
IT Engineer
Corporate Human Resources Team

Project Controls Department
Project Controls Manager
Planning Manager
Lead Scheduler
Scheduling Engineer
Budget Controller
Cost Engineer

Contract Department
Project Contract Manager
Contract Administration Manager
Subcontracts Manager
Contracts Engineer
Insurance Clerk

Engineering Department	
Engineering Manager	
Process Engineer	Project Information Manager
Mechanical Engineer	Information Coordinator
Electrical Engineer	Data Inputter
Equipment Engineer	IT Engineer
Piping Engineer	Engineering Data DCC
Instrument Engineer	Correspondence DCC
Civil Engineer	
Engineering Design Subcontractor(s)	

Procurement Department	
Procurement Manager	
International Buyer – Mechanical	Logistics Manager
International Buyer – Electrical	Expeditor
International Buyer – Pipes/Piping	Materials Manager
International Buyer – Instruments & Communications	Materials Controller
International Buyer – General & Miscellaneous	Shipping/Transport Clerk
Local Buyer	Warehouse Staff
Procurement DCC	Materials Tracking Assistant

Construction Department
Construction Manager
Mechanical Supervisor
Piping Supervisor
Pipeline Supervisor
Static/Rotating Equipment Supervisors
Electrical Supervisor
Instrument Supervisor
Civil Supervisor
Civil Engineer
On-Site Document Control Centre
Take-Over & Handover Teams
Camp Boss & Camp Team

Commissioning Department
Commissioning Manager
Operations Readiness Manager
Senior Process Engineer
Senior Instrument Engineer
Senior Instrument Lead
Senior Instrument Technician
Senior Electrical Lead
Senior Electrical Technician

HSE Department
HSE Manager
Security Manager
HSE Coordinator
HSE Officers
Security Coordinator
Security Personnel
Doctor
Medical Personnel

QA/QC Department
QA/QC Manager
Lead QA
QC Supervisor
Vendor Inspection Supervisor
Vendor Inspection Coordinator
General QC Inspectors
Coating Inspector
Welding Inspector

About the Author

The author of this book is Walter Alfred Salmon, a Chartered Quantity Surveyor. He has worked outside of the UK full time since 1980 and, prior to that, had spent a little time working in Holland and Nigeria. He commenced his career at a time when electro-mechanical calculators (such as the 'Anita' comptometer) were still relatively new and therefore expensive. As a result, and since ready-reckoners were banned, most arithmetical calculations were done longhand in those days, on paper. The dimensions on the drawings were still staunchly imperial (feet and inches), so the calculations had to be done using the duodecimal system (base 12). Subsequently he joined the company that pioneered the use of computers in quantity surveying (Monk & Dunstone), and he has been a champion for computerised systems in the construction industry ever since.

Mr Salmon's work experience since 1980 extends to managing Quantity Surveying and Project Management Consultancies in Qatar, Hong Kong and the Philippines. His involvement in handling the contractual and commercial side of major EPC work covered the Passenger Terminal No. 3 Project at Ninoy Aquino International Airport (Manila, the Philippines), the Bosphorus Crossing Tunnel and the Marmaray Commuter Rail Projects (Istanbul, Turkey), the South Sumatra to West Java Gas Pipeline Project (Indonesia), the Dung Quat Refinery Project (Vietnam), the Abu Dhabi Crude Oil Pipeline Project (the UAE), and the Rumaila Power Plant Project (Iraq).

Until very recently Mr Salmon was the Contracts Advisor to the General Manager of an international contracting organisation that specialises in carrying out EPC Projects for the Oil and Gas Industry in the Middle East, where he had been working for the past 10+ years. During the course of his career Mr Salmon was actively involved in risk management for a number of large-scale EPC projects, on a few of which he was heavily involved in preparing or defending major claims. In the course of this, he witnessed money being lost in many directions by the contractors he was involved with, most of which he considered were avoidable losses. Consequently, he developed a very keen interest in trying to identify precisely where EPC Projects were going wrong, and how best to ensure that the companies he worked for made the profit they had anticipated at the bidding stage. This book documents the primary lessons he learned about where the major risks sat for the EPC Projects he was involved with, and what to do about controlling similar risks on future projects, so that the profit anticipated at the bidding stage stands more chance of being realised.

Index

www.ingramcontent.com/pod-product-compliance
Lightning Source LLC
Chambersburg PA